The Astronomical Tables
of Giovanni Bianchini

History of Science and Medicine Library

VOLUME 12

Medieval and Early Modern Science

Editors

J.M.M.H. Thijssen, Radboud University Nijmegen
C.H. Lüthy, Radboud University Nijmegen

Editorial Consultants

Joël Biard, University of Tours
Simo Knuuttila, University of Helsinki
John E. Murdoch, Harvard University
Jürgen Renn, Max-Planck-Institute for the History of Science
Theo Verbeek, University of Utrecht

VOLUME 10

Presentation of Bianchini's tables to the Emperor, Frederick III (Ferrara, Biblioteca Comunale Ariostea, MS I.147, f. 1r).

The Astronomical Tables
of Giovanni Bianchini

By

José Chabás and Bernard R. Goldstein

BRILL

LEIDEN • BOSTON
2009

On the cover: Bianchini presents his Tables to Emperor Frederick III (Ferrara, Biblioteca Comunale Ariostea, MS I.147, f. 1r); detail from the frontispiece. See also Chapter One, p. 15.

This book is printed on acid-free paper.

Library of Congress Cataloging-in-Publication Data

Chabás, José, 1948–
   The astronomical tables of Giovanni Bianchini / by José Chabás and Bernard R. Goldstein.
      p. cm. — (History of science and medicine library, ISSN 1872-0684 ; v. 12)
   Includes bibliographical references and index.
   ISBN 978-90-04-17615-7 (hardback : alk. paper)
   1. Ephemerides—Europe—History—15th century. 2. Alphonsine Tables.
3. Bianchini, Giovanni, 15th cent. 4. Astronomy, Medieval. I. Goldstein, Bernard R.
II. Title. III. Series.

   QB7.C43 2009
   528—dc22

                                                                    2009010199

ISSN 1872-0684
ISBN 978 90 04 17615 7

PRINTED IN THE NETHERLANDS

CONTENTS

PREFACE

Astronomical tables abound in medieval manuscripts and early modern printed texts. Moreover, many of these documents, commonly found in libraries around the world, are restricted to tables, normally containing few words other than those in the titles and the headings of the tables. And it is not unusual to find manuscripts and printed texts with hundreds of pages composed of vast quantities of numbers, in the thousands, or even in the hundreds of thousands. Of course, the sizes of the tables vary substantially from short tables consisting of just two columns and a few rows, to monumental tables requiring tens of pages to reproduce them. Not only were certain sets of astronomical tables copied again and again by patient scribes or typesetters in the early days of printing, or by the astronomers themselves, but many of their users found ingenious ways to ameliorate them, generating new tables based on new approaches, new parameters, or that were just more user-friendly. The number of different sets of astronomical tables extant in manuscripts, produced in the Middle Ages and the early modern period, has never been evaluated but it would not be surprising that it is well beyond a thousand. Clearly, during this long period of time astronomical tables were a major way to convey astronomical ideas.

The authors of the present monograph have invested much time and effort in understanding and explaining astronomical tables that were produced in many countries, written in a variety of languages, and dealing with a great number of different astronomical issues. We are convinced that "cracking" an astronomical table, by bringing to the surface the model on which it is based and the parameters on which it relies, is a way to gain deep insight into how astronomy was conceived and practiced at the time and how it was transmitted to subsequent generations.

In examining hundreds of astronomical tables we have found some that greatly impressed us for the cleverness of their authors in grasping a problem that had been addressed by many previous astronomers and giving innovative solutions. We have also found tables that limited themselves to reproducing the astronomical tradition of their time with little innovation, and others where the main contribution consisted in facilitating the work of astronomical practitioners. And yet in many

cases the innovation was not in the models or the parameters underlying them, but in the approach to the various astronomical problems
to be solved.

The tables of Giovanni Bianchini (d. after 1469) are certainly voluminous; indeed, they are the largest set of astronomical tables produced
in the West before modern times, as far as we know. For many years
we cherished the idea of "cracking" these tables, but the task seemed
daunting, given their volume. Today we have reached our goal, and
our respect for Bianchini has shifted from volume to value. Although
not innovative in their building blocks, his tables reflect a well defined
approach to astronomy, a practical way to present it, and a solid computing ability. Now we can understand why the wealthy and powerful
d'Este family in 15th-century Ferrara engaged him to keep track of
their finances.

We thank the Biblioteca Comunale Ariostea of Ferrara for giving us
permission to reproduce an image in MS I.147, and all other libraries
that have made available to us the manuscripts and printed editions
mentioned in this monograph.

José Chabás – Bernard R. Goldstein
Rome – Pittsburgh
May 2008

LIST OF ILLUSTRATIONS

Frontispiece (opposite the title page). Presentation of Bianchini's tables to the Emperor, Frederick III (Ferrara, Biblioteca Comunale Ariostea, MS I.147, f. 1r)

INTRODUCTION

One of the characteristics of an exact science is to explain theoretically the processes addressed by that discipline and to treat them precisely in a quantitative way. As an exact science, astronomy makes extensive use of numerical computations. In early astronomy, this is best exemplified by astronomical tables in the tradition of Ptolemy's *Almagest* that dates to the 2nd century, A.D. Throughout the Middle Ages, astronomers compiled a great variety of astronomical tables to help computers determine the positions and other circumstances of the celestial bodies and to help them solve astronomical problems related to the daily rotation, the determination of the times of eclipses, etc. The simplest astronomical table consists of two columns of numbers, such that for each value in the first column there corresponds one and only one entry, representing what nowadays is called a function. The first column gives the successive values of the argument, currently called the independent variable. The information contained in such a table can be represented as a two-dimensional graph, although it was not done so at the time. Astronomical tables in the *Almagest* were not all of the simplest kind, for many of them had more than two columns, such that the entries in each column depended on a single argument. In the Latin West a special form of astronomical tables, introduced in the 14th century, is called a 'double argument table', that is, a table with two arguments, one set at the head of each column and another set at the head of each row, corresponding (in modern terms) to a function of two variables. This ingenious type of table can be represented as a three-dimensional graph. The advantage of double argument tables is that they reduce the number of steps in the computation of a planetary position, etc. But this meant that the table maker had to produce many more entries in the tables, which required him to perform a large number of computations. As far as we know, the earliest sets of double argument tables in the West were due to John of Murs (*c.* 1321), John of Lignères in Paris (*c.* 1325), and William Batecombe in Oxford (1348), all of whom based their tables on the Parisian Alfonsine Tables that began to circulate in the 1320s.

Astronomers addressing a variety of problems put together different tables and compiled 'sets of tables', that is, consistent collections
of astronomical tables embracing all or some aspects of mathematical
astronomy and usually accompanied by a text, called 'canons', explaining their use. Most sets of tables compiled in western Europe in the
Middle Ages followed the structure of those composed in Arabic, that
is, handbooks called zijes (from the Arabic *zīj*, plural *zījāt*).[1]

In the Renaissance the mathematical sciences played an important
role in humanistic culture, and they were highly appreciated at various social levels.[2] Mathematical astronomy was regarded as especially
valuable, for it was associated with cosmology and philosophy, as well
as with astrology and astrological medicine. In other words, there was
a considerable market for publications that included almanacs, ephemerides, and lunaria, several of them ranking among the best sellers
in scientific publications,[3] as well as sets of astronomical tables. The
investment of time and money by the printer in producing these sets
of tables was significant, and the fact that more than one edition of
the same set appeared is an indication of the popularity of this genre.
In particular, Bianchini's tables were printed three times between 1495
and 1553.

By the time Bianchini compiled his tables (*c.* 1442), European astronomers had access to several variants of the Parisian Alfonsine Tables, that
is, a set of tables which was recast beginning in the 1320s by a group of
notable scholars working in Paris, all of them sharing the name, John
(John of Murs, John of Lignères, and John of Saxony), and based on
the work done by the astronomers in the service of King Alfonso X of
Castile in the second half of the 13th century. Unfortunately, only the
canons of the original Castilian Alfonsine Tables are extant, not the
tables themselves.[4]

All these sets of astronomical tables are in the tradition of Arabic
zijes: they contain a great many tables and, at their core, are those for
the determination of the positions and motions of the five planets and
the luminaries. The position of a planet in longitude (along the ecliptic),

---

[1] See, e.g., Kennedy, *Survey.*
[2] See, e.g., Rose, *Renaissance of Mathematics.*
[3] See Chabás, *Granollachs.*
[4] Chabás and Goldstein, *Alfonsine Tables of Toledo.*

as well as the Sun and the Moon, is computed from tabulated values for mean motions and equations, that is

$$\lambda = \lambda_0 + \Delta\bar{\lambda} + c,$$

where λ_0 is the initial mean longitude of a planet at some epoch, $\Delta\bar{\lambda}$ is the increment in mean motion from epoch, t_0, to another time, t, and c is the equation, i.e., the deviation from mean motion to be computed from a model for planetary motion. In turn, $\Delta\bar{\lambda}$ is a linear function of time,

$$\Delta\bar{\lambda} = \mu \cdot (t - t_0),$$

where μ is the mean motion per day (or some other unit of time). In the case of the five planets, c depends on two variables, λ and α, the argument of anomaly, and

$$\alpha = \alpha_0 + \Delta\bar{\alpha} + c',$$

where α_0 is the argument of anomaly at epoch, $\Delta\bar{\alpha}$ is the increment in the argument of anomaly from t_0 to t, and c' is the equation of anomaly. To compute a position of one of the 5 planets using the tables, one first needs to find 3 quantities:

$$\bar{\lambda} = \lambda_0 + \Delta\bar{\lambda},$$
$$\bar{\kappa} = \lambda_A - \bar{\lambda},$$

where $\bar{\lambda}$ is the mean position of the planet and $\bar{\kappa}$, the mean argument of center, is the distance of $\bar{\lambda}$ from the planet's apogee, λ_A, and

$$\bar{\alpha} = \alpha_0 + \Delta\bar{\alpha}.$$

Then, with $\bar{\kappa}$ and $\bar{\alpha}$ as arguments in one or more tables,

$$\lambda = \bar{\lambda} + c(\bar{\kappa}, \bar{\alpha}).$$

Note the both c and c' can be positive or negative. These rules are common to all sets of astronomical tables in the Ptolemaic tradition, and they have generated a large number of tables based on various ingenious procedures. For worked examples of Bianchini's procedure for finding the true longitude of the Moon and of Mars, see Tables 17 and 41, below.

The Alfonsine tradition lies within the Ptolemaic tradition, having a set of parameters in common as well as the same underlying model. Nevertheless, we find a great variety of tables in this tradition, and among them are those of Bianchini, the subject of this monograph. As we shall see below, in addition to tables for planetary longitude, there are tables for eclipses, planetary latitudes (the distance of a planet north or south of the ecliptic at any given time), etc. But let us first consider some of the sets of tables produced in the 14th century.

The Parisian astronomers drew up tables and wrote their corresponding canons, based on the earlier Castilian material. One characteristic feature of the resulting Parisian Alfonsine Tables is the division of a circle into 6 physical signs of 60°, although in many cases the circle is divided into 12 zodiacal signs of 30°, as was the case in almost all previous sets of tables. These tables rapidly superseded the Toledan Tables, compiled at the end of the 11th century, also in Toledo, and extant in hundreds of manuscripts.[5]

Among the collections of tables ascribed to the Johns are (i) the tables of 1321, with a short canon, by John of Murs; (ii) the so-called *Tabule magne*, by John of Lignères; and (iii) the tables of 1322, with canons, also by John of Lignères. Unfortunately, a thorough study of these tables is still lacking.

(i) John of Murs compiled a set of tables to compute the positions of the planets and the luminaries which are extant in two MSS in Oxford and Lisbon.[6] In both of them physical signs of 60°, rather than zodiacal signs of 30°, are used. The tables are accompanied by a short text beginning *Si vera loca planetarum per presentes tabulas volueris invenire a tempore incarnationis domini dato perfecto deme 1320*... The presentation given to the tables of the planets is indeed original, because the organizational principle is a succession of mean conjunctions of each planet with the Sun. For each of the planets and the Moon we are given several tables: the first lists a number of mean conjunctions with the Sun, where the radix is 1321 (1320 completed); the second is a double argument table (30 × 30) to find the true positions of the planet, or the Moon, at times between two successive conjunctions; the third displays the mean longitude and the mean argument of center of the planet; the fourth lists the equation of center (the correction to be applied to the mean argument of center to obtain the true argument of center); and the fifth is a reduced

---

[5] Pedersen, *Toledan Tables*.
[6] Poulle, "John of Murs," 133.

double argument table (16 × 7) for the latitude of the planet. For Mercury and Venus, there is another column for the 3rd component of latitude (see Tables 73 and 74, below). All the parameters used here will later be found in the Parisian Alfonsine Tables (e.g., the maximum equations of center for Jupiter and Venus are 5;57° and 2;10°, respectively, thus departing from the parameters used in the Toledan Tables). It is most noteworthy that the titles of all tables other than those with double argument indicate that they were computed for Toledo, adding that it is 0;48h distant from Paris, and use the radices given by King Alfonso of Castile.

(ii) In 1322 (a date that appears in the colophon of the manuscripts) John of Lignères compiled another set of tables with canons in 44 chapters for the prime mover (*Cujuslibet arcus propositi sinum rectum…*), i.e., problems of trigonometry, the daily rotation, etc., and canons in 46 chapters for the motions of the planets and the determination of eclipses (*Priores astrologi motus corporum celestium…*). These canons have been edited, but not published.[7] Both canons describe one set of tables computed for the meridian of Paris which have December 31, 1320 as epoch; these canons and tables are extant in many manuscripts and the latter are often referred to as 'Alfonsine tables'.

(iii) The *Tabule magne* by John of Lignères were compiled in about 1325 and depend on the tables for 1322 by the same author, according to Poulle.[8] They are associated with the incipit, *Multiplicis philosophie variis radiis…*, and seem to be extant in only a few MSS. Curiously enough, the tables in Erfurt, MS F.388, use zodiacal signs of 30° whereas those in Lisbon, MS Ajuda 52–XII–35, ff. 67r–92v, use physical signs of 60°. There are four types of tables. The first gives the daily mean motions for a year for Saturn, Jupiter, and Mars (mean motions in longitude and mean arguments of anomaly); mean arguments of anomaly for Venus and Mercury; mean motions of the Sun, Venus, and Mercury, and the lunar nodes; mean motions in longitudes and mean argument of anomaly of the Moon. The Erfurt MS gives radices for the planets without indicating the place for which they are valid, but the values in the text indicate that the place is Paris and the epoch, the Incarnation (i.e., Jan. 1, 1 A.D.).

---

[7] Saby, *Jean de Lignères*.
[8] Poulle, "Alfonsine Tables and Alfonso X," 103.

The second displays the yearly mean motions of the above quantities from 1 year to 20 years (at intervals of 1y), from 20 years to 100 years (at intervals of 20y), and from 100 to 1000 years (at intervals of 100y). The third is for the planetary equations, presented as large double argument tables (31 × 60), and for the solar equation (with a maximum of 2;10°), giving a combined correction to be added or subtracted to the mean longitude of the planet to obtain its true position. The fourth presents the hourly true motion of the Moon. In the Erfurt MS the set continues with a voluminous table for the computation of the time from mean to true syzygy, known as *Tabulae Permanentes*, attributed to John of Murs.[9]

To these sets one should add the tables of John Vimond, an astronomer also working in Paris, who compiled tables with 1320 as epoch, for the use of students at the University of Paris.[10] These tables form a coherent set with all the elements needed to compute the positions of the celestial bodies, and are prior to, and independent of, the tables compiled in the early 14th century, which we call the Parisian Alfonsine Tables, also based on Castilian sources. Vimond's tables and the Parisian Alfonsine Tables have many parameters in common both for mean motions and equations.

These sets, and possibly those by other astronomers, gave rise to the Parisian Alfonsine Tables. This set underwent significant developments from the mid-14th century to the time of Bianchini, and at least three different adaptations were produced in three different countries.

(i) The Oxford Tables of 1348, often called *Tabule anglicane*, have been ascribed to William Batecombe and are associated with the canons beginning *Vera locum omnium planetarum in longitudine....*[11] The tables for mean motions give the mean argument of center rather than the mean longitude of the planet. There are also double argument tables for the planets, as was the case with the *Tabule magne* by John of Lignères and, according to North, this set was indeed the source for the double argument tables compiled in Oxford. In this case the entries are also given as a

---

[9] Chabás and Porres, "True Syzygies."
[10] Chabás and Goldstein, "Tables of John Vimond."
[11] North, "Alfonsine Tables in England."

function of the mean argument of center and the mean argument of anomaly, but the procedure to compute the true position of the planet is made easier than in the *Tabule magne*.[12] To illustrate the size of the double argument tables for the planets, it is perhaps sufficient to say that in the manuscripts they fill about 66 pages crowded with entries of about 39,600 numbers of one or two digits.

(ii) Around 1424 the Paduan astronomer Prosdocimo de' Beldomandi wrote canons and tables, based on Jacopo de Dondi's tables which, in turn, depended on the Parisian Alfonsine Tables.[13] The associated canons begin with *Facta et ordinata sunt quam plura et varia paria tabularum ad celestes motus*, and the tables contain all that is needed to compute positions of the celestial bodies. They also include a catalogue of more than 1,000 stars which served as the basis for the star catalogues published in the incunabula editions of the Parisian Alfonsine Tables. The presentation of Prosdocimo's mean motion tables follows the pattern in the Toledan Tables and the Tables of Novara (a version of the Toledan Tables adapted to the Christian calendar), using groupings of 28 years, rather than the presentation in the *editio princeps* of the Alfonsine Tables (1483), where one finds tables with multiples from 1 to 60 of the basic parameter (i.e., the daily mean motion) in each case (to be used with time intervals in days arranged in strict sexagesimal notation, rather than in years, months, days, and hours). On the other hand, Prosdocimo's tables for planetary latitudes with 22 columns are peculiar and unprecedented.

(iii) The Austrian astronomer, John of Gmunden (*c.* 1380–1442), compiled several versions of a set of astronomical tables for which he also wrote canons, and they were intended to accompany his lectures at the University of Vienna from 1419 onwards.[14] He collected many tables from his predecessors in the Alfonsine tradition and his set of astronomical tables (90 tables for the planets and the luminaries) is probably the most complete until that time. John of Gmunden reproduced earlier tables, adapted others, and modified the format of still others, but in all of them he was faithful to Alfonsine astronomy and did not depart from its parameters and

---

[12] North, "Alfonsine Tables in England," 278–82.
[13] Chabás, "Prosdocimo de' Beldomandi."
[14] Porres, *Jean de Gmunden.*

procedures. As is the case with many other sets of tables mentioned above, John of Gmunden's tables were never printed, but they were widely diffused in Central Europe for about a century.

Although these sets are not identical either in the number of tables they contain or in the format of some particular tables, they are organized in a way that is similar to the Parisian Alfonsine Tables. This was the state of the Alfonsine material in about 1442, when Giovanni Bianchini completed his voluminous set of tables. As we shall see, his set depends directly on the Alfonsine tradition, but differs from all previous sets in various crucial ways: the tables for the planets and the luminaries have a consistent format based on an internal organizing principle different from other sets of tables (see Tables 17, 27, 34, 41, 48, 56, 63, and 64, below), and his tables contain the largest number of entries ever computed in the Alfonsine tradition.

Bianchini did not use a single epoch in his tables, but a variety of 'intermediate' epochs from which all quantities involved in a computation are counted. So, for the Moon Bianchini chose to refer all quantities related to its motion to its mean anomaly, counted from the beginning of the most recent anomalistic month, whatever the date on which this occurs. Analogously for the planets, the intermediate epoch for each of them is the beginning of their respective most recent periods of anomaly. In the case of syzygies, all quantities are counted from the beginning of the most recent synodic month. This is a characteristic feature of Bianchini's tables, an arrangement which differs substantially from other sets of tables, even those sharing the same models and parameters, such as the Parisian Alfonsine Tables. In order to link the values obtained through the use of this scheme with dates expressed in years, months, and days, Bianchini introduced tables for computing a set of radices whose entries are associated with specific moments traditionally used by astronomers, or dates in his own time, that have to be understood as the time intervals between the beginning of an anomalistic month or the beginning of a period of planetary anomaly and the date in question. In the case of a mean synodic month, one has to compute the time interval between the beginning of a specific mean synodic month and the Incarnation. After finding these epochs, one then has the arguments for entering double argument tables (that require interpolation) from which one obtains the quantities to be added to the mean quantities at the intermediate epoch. The whole construction reflects a deep astronomical insight, but has the disadvantage of being unfamiliar and complex, which probably made some practitioners of astronomy avoid these tables, given the relative simplicity of the Parisian Alfonsine Tables and other adaptations of them.

Probably due to their substantial size and complexity, the tables of Bianchini were not copied very often in manuscript, but frequently enough to suggest to the printer that there was a market for them. And so they were published in 1495 in Venice for the first time. In the meantime, they coexisted with at least two other sets of tables that also addressed all problems related to the motions of the planets and the luminaries: the *Tabulae resolutae*, compiled and originally diffused in central Europe (first edited by A. Lacher, and published in Frankfurt in 1511 under the title *Tabulae resolutae de motibus planetarum aliorumque super celestium mobilium*), and the set of tables compiled by Abraham Zacut in Spain (first published in Leiria, Portugal, in 1496). Both of them presented the Alfonsine material in their own special way.

The first version of the *Tabulae resolutae* seems to be the work of Petrus Cruciferus, who compiled them for the meridian of Wrocław with 1424 as epoch. Various Polish astronomers, including Marcin Król, Andrzej Grzymała, Albert of Brudzewo, and John of Glogovia expanded them, resulting in a set of tables in the Alfonsine tradition that had a wide diffusion in manuscript in the 15th century and then in print in the 16th century.[15] A typical set of the *Tabulae resolutae* contain tables for the radices and mean motions of 12 quantities: the apogees and the fixed stars; access and recess of the 8th sphere; Sun, Venus, and Mercury; Moon; argument of the Moon; argument of latitude of the Moon; lunar node; Mars; Jupiter; Saturn; argument of Venus; and argument of Mercury. The entries are arranged according to a system of cyclical radices *ad annos collectos* at 20-year intervals. There are also tables for mean syzygies, listing four variables: time of mean syzygy, mean motion of the Moon, mean argument of the Moon, and mean argument of latitude of the Moon. Then follow tables for the positions and motions of the apogees of the planets, also tabulated at intervals of 20 years; tables for the equations; interpolation tables; tables for the daily motions of the Sun and the Moon, the equation of time, and the planetary stations and retrogradations; and tables for the rising times. There are no tables with double arguments in any of these versions.

---

[15] Dobrzycki, "The *Tabulae Resolutae*"; Chabás, "Astronomy in Salamanca"; Chabás, "Diffusion of the Alfonsine Tables."

The *Tabulae resolutae* do not present any innovation in the parameters underlying the tables, for all are strictly in the Alfonsine tradition, but the material is presented differently from that in previous tables, possibly making them easier to use. They were printed several times—even as late as 1588—thus competing for a few decades with the Prutenic Tables, first printed in 1551, based on Copernicus's theories.[16]

The Great Composition (*ha-Ḥibbur ha-gadol*) is the set of tables compiled in Hebrew by Abraham Zacut (1452–1515) in Salamanca.[17] It consists of about 65 tables and lengthy canons explaining their use. The tables have 1473 as epoch, they are arranged for the Christian calendar, and their entries are computed for the meridian of Salamanca. The work was finished around 1478 and three years later it was translated from Hebrew to Castilian. In the *Ḥibbur* there are some double argument tables, in particular for the latitudes of the five planets and for the unequal motion of Mercury. Zacut adheres to the Alfonsine tradition, which came to Salamanca in the form of the *Tabulae resolutae*, as well as to the astronomical tradition in Hebrew, which included works by Levi ben Gerson, Jacob ben David Bonjorn, and Judah ben Asher II (all 14th century). Two versions of Zacut's *Ḥibbur* (one with canons in Latin and the other with canons in Castilian), edited by the Portuguese scholar Joseph Vizinus, were published in 1496 in Leiria, Portugal, under the title *Tabulae tabularum coelestium motuum sive Almanach Perpetuum*. Shortly thereafter, in 1498, this work was reprinted in Venice, with three more editions appearing during the 16th century.

Thus, in the early years after the invention of printing with movable type, the Alfonsine corpus was well established in at least seven European countries, with no other set of tables to compete seriously with it. Printing definitely accelerated the diffusion of the various presentations of the Alfonsine Tables throughout Europe, but the first set of tables to be published was a form of the Parisian Alfonsine Tables. Before 1500 there were two editions which differ in some important ways from one another, printed in the same city:

> 1483 – *Tabule astronomice illustrissimi Alfontij regis castelle.* Venice: Erhard Ratdolt;

---

[16] Reinhold, *Prutenicae tabulae coelestium motuum.*
[17] Chabás and Goldstein, *Abraham Zacut.*

1492 – *Tabule Astronomice Alfonsi Regis.* Venice: Johannes Lucilius San-
    tritter.

Then followed the first edition of Bianchini's tables, also printed in
Venice:

1495 – Giovanni Bianchini. *Tabulae astronomiae.* Venice: Simon Bevi-
    laqua.

Before the turn of the century, however, another set of tables, the *Almanach Perpetuum*, came out in two editions, as mentioned above:

1496 – Abraham Zacut. *Almanach Perpetuum.* Leiria: Samuel d'Ortas;
1498 – Abraham Zacut. *Ephemerides sive Almanach Perpetuum.* Venice:
    Johannes Lucilius Santritter.

The *editio princeps* (1483) of the Parisian Alfonsine Tables devotes 120
pages to numerical tables, excluding the star catalogue. In these pages
we have counted more than 51,000 numbers of one or two digits, of
which 41,000 have been computed, and about 10,000 are for the arguments in the tables (i.e., belonging to sets of consecutive numbers). The
tables authored by Bianchini fill 633 pages in the edition of 1495 and
contain about 315,000 numbers of one or two digits, that is, more than
6 times the amount in the Parisian Alfonsine Tables. Of these, 300,000
have been computed (more than 7 times the amount in the Parisian
Alfonsine Tables), and about 15,000 are for the arguments. It should
be noted that the second edition of Bianchini's tables (1526) enlarged
the number of tables from 68 to 111, thus considerably increasing this
huge undertaking of printing numbers.

Bianchini was not the only one in the 15th century to produce planetary tables with a vast quantity of entries to be published later in the
century by courageous printers. In the *Almanach Perpetuum*, printed in
Leiria in 1496, with 306 pages of numerical tables (excluding the star
list), we have counted more than 201,000 numbers, of which 187,000
have been computed (4.5 times the amount in the Parisian Alfonsine
Tables), and about 14,000 are for the arguments. It is thus clear that
Bianchini and, to a lesser extent, Zacut made enormous efforts to provide their readers with a lot of precise numerical information presented
as astronomical tables that could help them compute accurately the
positions of the luminaries and the five planets. In rendering this task
easier, they surely increased the number of practitioners of astronomy,
a purpose which was greatly facilitated by the use of printing (or even
inconceivable without it).

CHAPTER ONE

GIOVANNI BIANCHINI: LIFE AND WORK

What little is known about the life of Giovanni Bianchini (Latin: Iohannes Blanchinus) was reported by G. Federici Vescovini (1968). Bianchini was probably born in the first decade of the 15th century, and he died sometime after 1469. His family came from Florence, but his father, Amerigo, was established in Bologna by 1400. Bianchini was a merchant in Venice until 1427 and then worked for Nicolò d'Este (1383–1441), Marquis d'Este, Signore of Ferrara, Modena, Parma, and Reggio. Bianchini spent most of his life in Ferrara where, for about three decades, he served as administrator of the estate of the prominent d'Este family, first for Nicolò, and then for Leonello (1407–1450) and Borso (1413–1471). Bianchini also taught at the University of Ferrara,[1] and it has been established that he visited other Italian cities including Milan, Venice, Bologna, and Rome.[2] The date of his death is uncertain. It is known, however, that he was buried at Saint Paul's Church in Ferrara, which was destroyed by an earthquake in 1570.

Bianchini's scientific writings, all of which deal with astronomy and mathematics, were composed between 1440 and 1460. In 1463–1464 he corresponded with Regiomontanus (1436–1476). The five extant letters (two from Bianchini and three from Regiomontanus) have been edited repeatedly and much has been written about them.[3] The letters mainly concern astronomical and mathematical problems and their solutions. The first extant letter is dated July 27, 1463: it was written by Regiomontanus, then in Venice, in reply to a letter from Bianchini in late June 1463 that does not survive. At that time Regiomontanus was in Venice accompanying Cardinal Johannes Bessarion (1395–1472) and had intended to visit Bianchini in Ferrara shortly before his arrival there, but the rumor that plague was ravaging Ferrara prevented him

---

[1] Thorndike, "Bianchini in Paris Manuscripts," 5.

[2] Magrini, "Joannes de Blanchinis Ferrariensis," 8.

[3] Murr, *Memorabilia*; Curtze, "Briefwechsel Regiomontan's"; Gerl, *Briefwechsel Regiomontanus–Bianchini*; Zinner, *Regiomontanus, his Life and Work*, 60–69; Swerdlow, "Regiomontanus on Critical Problems."

from meeting Bianchini. Most historians think it unlikely that Bianchini and Regiomontanus ever met.

In this monograph we focus on Bianchini's best known work, his *Tabulae astronomiae* (1495), consisting of astronomical tables and a set of canons explaining their use. We will demonstrate that these tables depend on the Alfonsine Tables that were later printed in Venice in 1483 and 1492 (with extant manuscripts as early as the 14th century), but Bianchini added many special features which gave his tables a completely different presentation from that of the standard Alfonsine tables. The work was compiled in Ferrara and addressed to his patron, Leonello d'Este, probably in 1442.[4] There follows a list of the manuscripts, with the sigla we assign to those that we cite, that are known to include this set of tables (or parts of it):

> Cracow, Biblioteka Jagiellońska, MSS 555, 557, 603, and 606
> Ferrara, Biblioteca Comunale Ariostea, MS I.147
> Florence, Biblioteca Laurenziana, MS Pl. 29,33
> Milan, Biblioteca Ambrosiana, MSS C. 207 inf. and C. 278 inf.
> Naples, Biblioteca Nazionale, MSS VIII.C.34 [Na] and VIII.C.36 (only a few tables)
> Nuremberg, Stadtbibliothek, Cent V 57 [Nu]
> Oxford, Bodleian Library, MS Canon. Misc. 454
> Paris, Bibliothèque nationale de France, MSS 7269, 7270, 7271, and 16212
> Rome, Biblioteca Casanatense, MS 1673 [Rc]
> Vatican, Biblioteca Apostolica, MS Pal. lat. 1375 [Va]
> Venice, Museo Correr, MS Cicogna 3748
> Vienna, Österreichischer Nationalbibliothek, MSS 2293 and 5299

Va, ff. 1ra–8va, contain the Tables of Bianchini, heavily annotated by Johannes Virdung of Hassurt, as indicated on f. 1r. The tables themselves occupy ff. 185r–262v, and on their first page we are told that they were copied in Cracow in 1488 from June 12 to July 20.

In a letter dated 1456 to Johann Nihil, court astrologer to Emperor Frederick III, Georg Peurbach (1423–1461) described the calculation of ephemerides he had computed, together with Regiomontanus, for which they used Bianchini's tables.[5] While in Vienna in 1460 Regiomontanus made a copy of these tables for himself and wrote an abridged

---

[4] For the contents of some of the manuscripts that contain this work, see Boffito, "Tavole astronomiche di Bianchini"; Thorndike, "Bianchini in Paris Manuscripts"; Thorndike, "Bianchini in Italian Manuscripts"; and Rosińska, *Scientific Writings*.

[5] Hellman and Swerdlow, "Peurbach (or Peuerbach), Georg," 474.

set of canons for them entitled, *Canones breviati in tabulas Ioannis de Blanchinis*, beginning with *Augem planetarum comunem invenire* (Nu, ff. 5r–19v).[6]

In addition to the *edito princeps* of 1495, noted above, two editions of the canons and the tables appeared in the 16th century:

> 1526 – *Tabule Joa[nni] Blanchini Bononiensis*. Venice: Luca Antonio Giunta;
> 1553 – *Luminarium atque planetarum motuum tabulae octoginta quinque*. Basel: Joannes Hervagium.

In the edition of 1553 the authors are given as Giovanni Bianchini, Nicholaus Prugner, and Georg Peurbach. Indeed this volume contains two different works: Peurbach's *Tabulae eclipsium* consisting of canons and tables, and the Tables of Bianchini, edited by Prugner, which include a considerable number of tables compiled by the editor himself. The two earlier editions only contain Bianchini's work.

The book has two dedications, one to Leonello d'Este (d. 1450), opening with *Consideranti mihi, dive Leonelle,…*, and a later one addressed to Frederick III, Holy Roman Emperor (reigned: 1452–1493), beginning with *Cum nuper maiestas tua, Serenissime Cesar…* According to Magrini,[7] the dedication to Emperor Frederick was suggested by Bianchini's patron at the time, Borso, and was presented together with the tables to the Emperor on the occasion of his visit to Ferrara in January 1452. A splendid picture of that ceremony in the form of a miniature, ascribed to Giorgio d'Alemagna,[8] in a manuscript currently in Ferrara shows Bianchini kneeling before the Emperor, handing him his set of tables, and receiving from him his coat of arms. He is accompanied by Borso d'Este (see Frontispiece, opposite the title page).

Prior to the dedications in the first two printed editions, we find an encomium in praise of Bianchini's book that was written by Augustinus Moravus in January 1495 in Padua. In these editions the tables are preceded by canons consisting of an Introduction and 51 chapters. In contrast, MSS Na, Rc, and Va only include the dedication to Leonello d'Este, after which Bianchini's canons begin, consisting of an Introduc-

---

[6] For the dependence of various tables printed by Regiomontanus on Bianchini's *Tabulae astronomiae*, see Rosińska, "'Fifteenth-Century Roots' of Modern Mathematics", 67, n. 16.

[7] Magrini, "Joannes de Blanchinis Ferrariensis," 10.

[8] Medica, "Miniato da Giorgio d'Alemagna."

tion and 39 chapters (numbered consecutively in Na, but unnumbered in
Va) or 53 chapters in two parts of 41 and 12 chapters in Rc. The folios
in Na were not properly bound, with the result that the introduction
is found on f. 1r and f. 11r–v, for folios 2r–v and 11r–v have been
interchanged.

The Introduction begins with references to Ptolemy and the *Almagest*:
"Ptholomeus qui merito illuminator divine artis astrologie vocari potest,
in suo libro Almagesti...". Several scholars other than Ptolemy are
mentioned (Hipparchus, Thābit ibn Qurra, al-Battānī, the compilers of
the Toledan Tables, and Alfonso X, among others), but we note that
none of them is later than Alfonso (d. 1284). The Introduction focuses
on two astronomical matters, precession/trepidation and the latitude
of the planets. Bianchini praises the work of Alfonso and, as will be
seen below, in the first table in his treatise he addresses the problem of
precession/trepidation strictly in accordance with the approach taken in
the Alfonsine corpus. As for the planetary latitudes, Bianchini indicates
that he compiled tables following the instructions given by Ptolemy in
Book XIII of the *Almagest*, which is indeed the case, to overcome the
"significant discrepancy from the truth, especially for Venus and Mercury," found in other sets of tables.[9] The Introduction closes with some
basic information helpful to the reader when using his tables: years
are 365;15 days; the beginning of the year is March 1; the epoch is the
Incarnation; physical signs of 60° are used, as in the Parisian Alfonsine
Tables; computed examples and some tables are for Ferrara whose geographical coordinates are longitude 32° and latitude 45°; and motions
are referred to the 9th sphere (i.e., the coordinates are tropical).

The canons, that is, the Introduction followed by 51 chapters, are
the same in both editions (1495 and 1526), but for minor variants: in
ed. 1495, Chapters 1, 2, and 3 are not numbered, and in ed. 1526 the
chapter following 50 is numbered 49. The chapters in ed. 1495 are the
following:

[1.] De modo operandi per tabulas Joannis Blanchini generaliter ad
     quemcumque meridianum volueris.
[2.] Ad sciendum numerum dierum a principio anni ad quemcumque
     diem cuiusque mensis sequentis
[3.] Locum augium comunium octave spere in nona invenire

---

[9] Bianchini, *Tabulae astronomiae* (1495), a4v.

4. Ad inveniendum locum augis cuiuslibet planete
5. Medium motum solis per tabulas invenire volueris
6. Applicationem solis ad eius augem invenire
7. Ad idem per alias tabulas
8. Per augem solis auges aliorum planetarum invenire
9. Verum locum solis invenire
10. Introitum solis in ariete seu in alio signo vel gradu examinare
11. De examinatione cursus lune
12. Latitudinem lune per tabulas invenire
13. Verum locum capitis et caude draconis invenire
14. Vera loca planetarum trium scilicet superiorum, Veneris etiam atque Mercurii per tabulas invenire
15. De latitudine planetarum
16. De elongatione planetarum a sole et e converse
17. Radices Christi cuiuslibet planete ad quodlibet aliud tempus extendere
18. Exemplum ad inveniendum verum locum planetarum in longitudine et latitudine atque eorum distantiam a sole
19. Utrum planeta fuerit stationarius directus sive retrogradus
20. Utrum planeta fuerit orientalis sue occidentalis a sole
21. Tempus vere coniunctionis et oppositionis luminarium per tabulas Jo. Blan. invenire
22. Duodecim coniunctiones immediate sequentes faciliter invenire
23. Verum locum solis per totum annum velociter extendere
24. Verum locum lune per totum annum invenire
25. Vera loca trium superiorum, Veneris etiam atque Mercurii continuando extendere
26. Ad reducendum tempus calculi facti diebus non equatis ad dies equatos et postmodum ad horas horologii initius sexti climatis ad meridianum Ferrarie
27. Gradum ascendentem et subsequenter figuram duodecim domorum celi per tabulas Joannis Blanchini erigere
28. Nota admirabilem operationem per locum solis loca aliorum planetarum et eorum latitudinem per tabulas Jo. Blanchini invenire
29. Planetarum calculum per tabulas Jo. Bla. ad Alphonsii regulas reducere
30. De tribus inferioribus
31. Medium motum et argumentum solis per tabulas cuiusvis planetarum superiorum, Veneris quoquem et Mercurii invenire

32. Latitudinem planetarum et eorum centra et argumenta equata per tabulas invenire
33. De latitudine trium superiorum
34. De latitudine Veneris et Mercurii
35. De reformatione tabularum radicum planetarum
36. De examinatione tabularum lune
37. De examinatione tabularum trium superiorum, Veneris quoque et Mercurii
38. De elongatione planete ad solem et e converso
39. Ad inveniendum precise dies et horas stationum planetarum
40. Tempus revolutionum annorum mundi seu nativitatum ac etiam cūiuscumque alius principii punctaliter reperire
41. Gradum ascendentem et ceterarum domorum cuspides tam in radice alicuius principii quam in revolutionc suorum annorum perscrutari
42. Vera loca planetarum in quacumque revolutione velociter indagare
43. Verum locum capitis in revolutionibus annorum invenire
44. Feriam cuiuscumque mensis latinorum literam dominicalem aureum numerum et indictionem in quolibet anno a nativitate Christi indagare
45. De festis mobilius inveniendis
46. De generali doctrina operationis tabularum Jo. Blan. ad quemcumque calculum volueris in motibus planetarum que dicitur corona tabule
47. Utrum planeta sit auctus vel diminutus numero aut per calculum
48. Utrum planeta sit ascendens vel descendens in circulo sue augis
49. Exemplum de universali atque oportuno calculo tam in electionibus quam in interrogationibus nativitatibus annorum revolutionibus ac etiam de applicatione alicuius planete in aliquod signum vel gradum signi, que omnia ex compositione tabularum aperte demonstrantur
50. Moram nati in utero materno atque verum gradum ascendentem cuiuscumque nativitatis per tabulas Jo. Blan. perscrutari
51. De regulis multiplicandi atque dividendi in calculo

In the edition of 1526 the editor, Luca Gaurico, added 8 short paragraphs after Chapter 51. The title of the first is "Verum ascendentis gradum per earum seminis rectificare secundum Jacobum Dundum patavinum", apparently based on a previous work by the astronomer

Jacopo de Dondi of Padua (1298–1359), and that of the eighth, "Lati-tudinem 5 errantum supputare". On the other hand, ed. 1553 has only the first 18 chapters.

The tables come after the canons and have the same general title in the editions of 1495 and 1526, and a similar title is found in ed. 1553: *Tabule Ethereorum Motuum Secundi videlicet mobilis: Luminarum at Planetarum viri perspicacissimi Joannis Blanchini. Omnium ex his que Alfonsum Sequuntur quem facillime. Sidere felici Incipiunt.* Bianchini uses physical signs of 60° throughout this work but, curiously enough, not for the lunar anomaly, which he systematically gives in degrees from 0° to 360°.

In addition to his *Tabulae astronomiae*, Bianchini wrote a few other treatises:

1. *Compositio instrumenti.* This short treatise, dated 1442, addresses the construction and use of an instrument, called *biffa*, to determine the altitudes of the stars, and in it Bianchini explains the meaning and use of the decimal point. The text begins *Primo composui duas figuras (rigas) equalis longitudinis...*, and has been published.[10]
2. *Canones tabularum super primo mobile.* This treatise includes both text and tables devoted to spherical trigonometry. The text begins with *Non veni solvere legem sed illuminare his qui in tenebris sedent...* and gives two basic parameters: 40;45,4° for the latitude of Ferrara and 23;30,30° for the obliquity of the ecliptic.
3. *Flores Almagesti.* Bianchini's largest work, composed between 1440 and 1455, consists of 8, 9, or 10 treatises (the number of treatises varies in different manuscript traditions). The first treatise, *Arithmetica*, beginning *Arithmetica dico quod determinatur per numeros...*, deals with theoretical arithmetic and the resolution of numerical problems, and the second, called *Arithmetica algebrae* or *De algebra*, is devoted to the resolution of quadratic equations.[11] They were written around 1440 in Ferrara and, together with the third treatise, *De proportionibus*, serve as a mathematical introduction to Bianchini's astronomy.[12] The other treatises are directly concerned with astronomical matters and follow Ptolemy's *Almagest*, but do not go beyond *Almagest* VI.

---

[10] Thorndike and Kibre, *Catalogue of incipits*, col. 1099; Garuti, "Compositio instrumenti."

[11] Rosińska, "Euclidean *spatium*," 29.

[12] Rosińska, "Bianchini's *De Algebra.*"

4. *Canones tabularum de eclipsibus luminarium*. In this treatise, written between 1456 and 1460, Bianchini reports observations that he
made of four lunar eclipses that took place in 1440, 1448, 1451, and
1455,[13] as well as the computation of a solar eclipse to be observed
at Ferrara in July 1460.[14] The incipit is: *In libro florum Almagesti per
me Ioannem blanchinum demostratum est....*

5. *Tabulas magistrales*. This is a set of 6 or 7 tables, depending on the
manuscripts, arranged in two groups. Among them are high precision tables for tangents and cosecants, where Bianchini abandons
sexagesimal notation, replacing it with decimal notation, that is,
defining the radius of the circle as $R = 10^n$ (with n = 3 for tangents
and n = 4 for cosecants). Bianchini was the first mathematician in
the West to use purely decimal tables of trigonometric functions,
soon followed by Regiomontanus.[15]

In spite of their unusual presentation and lack of user-friendliness,
Bianchini's tables were appreciated and used by quite a number of
astronomers. To the names of his contemporaries, Peurbach and
Regiomontanus, who have already been mentioned, we may add that
of Alessandro Borromei (*c.* 1463), doctor of arts and medicine, who
lived in Venice.[16] Moreover, the copy of Bianchini's tables owned and
annotated by Johannes Virdung of Hasfurt is now preserved in MS
Va. In Cracow, Bianchini's work seems to have been especially well
regarded, for several manuscripts preserve copies of his tables while
others give evidence that they were used. At least two of the manuscripts were copied by Ioannes Zmora de Leśnicz, one in Perugia in
1453 (Cracow, MS 555) and the other in Cracow in 1456 (Cracow, MS
557). In another manuscript at the Biblioteka Jagiellońska, MS 1841, we
find a long text (ff. 9–65) beginning *Volo invenire verus motus solis pro
meridiano cracoviensi ex Blankini tabulas*, in 20 chapters explaining the
use of Bianchini's tables and giving examples of computations for July
1447.[17] Moreover, Biblioteka Jagiellońska, MS 2480, f. 140, contains tables
based on those of Bianchini and computed for 1459; and MS 2478, ff.
96–105, contains *Proposiciones in Tabulas magistri Iohannis Blankijni*.

---

[13] Thorndike, "Bianchini in Paris Manuscripts," 170.
[14] Magrini, "Joannes de Blanchinis Ferrariensis," 25.
[15] Rosińska, "Universities in Copernicus' Time."
[16] Magrini, "Joannes de Blanchinis Ferrariensis," 16.
[17] See also Rosińska, *Scientific Writings*, 457.

As noted by Rosińska: "In Cracow, Bianchini's mathematical (trigono-metric) and astronomical tables were systematically copied, rearranged, and adapted to Cracow meridian, to begin with Martinus Rex' students and to finish with Copernicus, and with his younger colleagues in Cra-cow."[18] On the other hand, the edition of 1526 we have consulted at the Library of the University of Barcelona indicates on its front page that this copy belonged to the Franciscan monastery of Barcelona "for the use of Jo. Salom", most probably the Franciscan priest who published a proposal in 1576 addressed to Pope Gregory XIII for correcting the Roman calendar.[19] There was also some diffusion of Bianchini's canons and tables in Hebrew.[20]

In the preceding paragraph, we have not attempted to provide a comprehensive study of the reception of Bianchini's tables; rather, our goal has only been to demonstrate that his work was not neglected by his contemporaries and immediate successors.

---

[18] Rosińska, "Universities in Copernicus' Time," 10.
[19] Ziggelaar, "Papal Bull of 1582," 205–6.
[20] See, e.g., Steinschneider, *Die hebraeischen Uebersetzungen*, 626–28; Steinschneider, *Die hebraeischen Handschriften in Muenchen*, 17; and Chabás and Goldstein, *Abraham Zacut*, 22.

CHAPTER TWO

ANALYSIS OF THE TABLES

1. Introduction

In the absence of an autograph copy, it is not possible to specify exactly the contents of the astronomical tables of Bianchini. Each manuscript copy and each edition we have consulted includes tables some of which may not be due to him. The problem is not unique to this case, for it applies to other sets of medieval tables that exist in multiple copies, such as the Toledan Tables or the Parisian Alfonsine Tables. As a living text, each copy was 'personal' for some astronomer and differs from copies in the possession of other astronomers, and each printed version was the result of personal decisions made by the editor or the printer. The *editio princeps* (1495) of Bianchini's tables furnishes a good illustration of this. On the very last page of the printed book, f. D6r, the editor adds a paragraph which reads:

> Note that some tables have deliberately been omitted, in particular those for the equations and latitudes...for they seem to be there [among Bianchini's original tables] almost without reason and superfluously, because the tables for the equations and latitudes of the planets are not uniquely found in Bianchini's own tables but have already been in circulation for some time, partly in printed form in the case of Alfonso and partly written by others....

The editor was clearly aware of the publication of the Alfonsine Tables in 1483 and 1492, both published in Venice (as was the first edition of Bianchini's tables), and he refers specifically to the tables for equations and latitudes that we have numbered 69–80, below. We offer the following classification to convey a sense of the diversity in the tradition of the manuscripts and printed editions:

(1) Tables 1 to 68 are found in ed. 1495 and all but one of them appear in Na, a good example of a copy of these tables in manuscript—the exception is Table 65;

(2) Tables 69 to 86 are found in manuscripts but not in ed. 1495; and

(3) Tables 87 to 112 are found in ed. 1526 but not in ed. 1495 or in Na.

The edition of 1526 also includes all tables in the edition of 1495, but those in category (3) are unlikely to be part of Bianchini's legacy. Regiomontanus made his own copy of Bianchini's tables, extant in Nu; it is certainly an early witness to Bianchini's work, but it is not the result of mere copying, for Regiomontanus omitted some of the tables (e.g., those for syzygies), abridged others, and modified still others. Rc is an example of the variety of tables collected together in the 'Tables of Bianchini', for it contains tables in all three of the categories listed above. Although we are aware of many other copies in manuscripts of the 15th century (and later), we do not report the variants in them; this would be a massive undertaking and is unlikely to shed much light on the astronomical content of these tables.

Distribution of tables in Na, Nu, Rc, Va, ed. 1495, and ed. 1526

	Title	Na	Nu	Rc	Va	1495	1526
1	Precession	•	•	–	•	•	•
2	Radices of the apogees	•	•	–	•	•	•
3	Days in a year beginning in January	•	•	–	•	•	•
4	Mean motion of the Sun and the lunar node in years	•	•	–	•	•	•
5	Radices of the mean Sun and the lunar node	•	•	–	•	•	•
6	Mean motion of the Sun and the lunar node in days	•	•	–	•	•	•
7	Mean motion of the Sun and the lunar node in hours	•	•	–	•	•	•
8	Position of the solar apogee	•	•	–	•	•	•
9	Planetary apogees	•	•	–	•	•	•
10	Days in a year beginning in June	•	–	–	•	•	•
11	True motion of the Sun	•	•	–	•	•	•
12	Mean motion of the Moon in collected years	•	•	–	•	•	•
13	Mean motion of the Moon in expanded years	•	•	–	•	•	•
14	Multiples of the mean anomalistic month	•	•	–	•	•	•
15	Radices of the Moon	•	•	–	•	•	•
16	Monthly mean motion of the Moon for a common year	•	•	–	•	•	•
17	True position of the Moon	•	•	–	•	•	•
18	Daily mean motion of the Moon	•	–	–	•	•	•
19	Hourly mean motion of the Moon	•	–	–	•	•	•
20	Lunar latitude	•	•	–	•	•	•
21	Mean motion of Saturn in collected years	•	•	–	•	•	•

Table (*cont.*)

	Title	Na	Nu	Rc	Va	1495	1526
22	Mean motion of Saturn in expanded years	•	•	−	•	•	•
23	Radices of Saturn	•	•	−	•	•	•
24	Periods of anomaly of Saturn	•	•	−	•	•	•
25	Daily mean motion of Saturn	•	−	−	•	•	•
26	Hourly mean motion of Saturn	•	−	−	•	•	•
27	True position of Saturn	•	•	•	•	•	•
28	Mean motion of Jupiter in collected years	•	•	−	•	•	•
29	Mean motion of Jupiter in expanded years	•	•	−	•	•	•
30	Radices of Jupiter	•	•	−	•	•	•
31	Periods of anomaly of Jupiter	•	•	−	•	•	•
32	Daily mean motion of Jupiter	•	−	−	•	•	•
33	Hourly mean motion of Jupiter	•	−	−	•	•	•
34	True position of Jupiter	•	•	•	•	•	•
35	Mean motion of Mars in collected years	•	•	−	•	•	•
36	Mean motion of Mars in expanded years	•	•	−	•	•	•
37	Radices of Mars	•	•	−	•	•	•
38	Periods of anomaly of Mars	•	•	−	•	•	•
39	Daily mean motion of Mars	•	−	−	•	•	•
40	Hourly mean motion of Mars	•	−	−	•	•	•
41	True position of Mars	•	•	•	•	•	•
42	Mean motion of Venus in collected years	•	•	−	•	•	•
43	Mean motion of Venus in expanded years	•	•	−	•	•	•
44	Radices of Venus	•	•	−	•	•	•
45	Periods of anomaly of Venus	•	•	−	•	•	•
46	Daily mean motion of Venus	•	−	−	•	•	•
47	Hourly mean motion of Venus	•	−	−	•	•	•
48	True position of Venus	•	•	•	•	•	•
49	Mean motion of Mercury in collected years	•	•	−	•	•	•
50	Mean motion of Mercury in expanded years	•	•	−	•	•	•
51	Radices of Mercury	•	•	−	•	•	•
52	Periods of anomaly of Mercury	•	•	−	•	•	•
53	Monthly mean motion of Mercury in anomaly	•	•	−	•	•	•
54	Daily mean motion of Mercury	•	−	−	•	•	•
55	Hourly mean motion of Mercury	•	−	−	•	•	•
56	True position of Mercury	•	•	•	•	•	•

Table (*cont.*)

	Title	Na	Nu	Rc	Va	1495	1526
57	Mean motions for conjunction in collected years	•	–	•	–	•	•
58	Radices of the Moon for conjunction	•	–	•	–	•	•
59	Multiples of half a mean synodic month	•	–	•	–	•	•
60	Mean motions for conjunction in expanded years	•	–	•	–	•	•
61	Monthly mean motions for conjunction	•	–	•	–	•	•
62	Time of mean conjunctions for a common year	•	–	•	–	•	•
63	True position of the Sun at mean conjunction	•	–	•	–	•	•
64	True position of the Moon at mean conjunction	•	–	•	–	•	•
65	Multiplication table	–	–	•	–	•	•
66	Calendaric table for collected years	•	•	•	–	•	•
67	Calendaric table for expanded years	•	•	•	–	•	•
68	Movable feasts	•	•	•	–	•	•
69	Solar equation	•	–	–	•	–	–
70	Latitude of Saturn	•	–	•	•	–	•
71	Latitude of Jupiter	•	–	•	•	–	•
72	Latitude of Mars	•	–	•	•	–	•
73	Latitude of Venus	•	–	•	•	–	•
74	Latitude of Mercury	•	–	•	•	–	•
75	Lunar equation	•	–	•	•	–	•
76	Equation of Saturn	•	–	•	•	–	•
77	Equation of Jupiter	•	–	•	•	–	•
78	Equation of Mars	•	–	•	•	–	•
79	Equation of Venus	•	–	•	•	–	•
80	Equation of Mercury	•	–	•	•	–	•
81	Stations	•	–	•	•	–	•
82	Equation of time	•	–	•	•	–	•
83	Length of daylight for Ferrara and Bologna	•	–	•	•	–	•
84	Domification for the 5th climate	•	•	•	–	–	•
85	Domification for the 6th climate	•	•	•	–	–	•
86	Domification for the 7th climate	•	•	Rc	–	–	•
87	Geographical coordinates	–	–	–	–	–	•
88	Solar equation and solar velocity at mean conjunction	–	–	–	–	–	•
89	Lunar equation and solar velocity at mean conjunction	–	–	–	–	–	•

Table (*cont.*)

	Title	Na	Nu	Rc	Va	1495	1526
90	Excess of revolution for the Sun in expanded years	–	•	•	–	–	•
91	Equation of the excess of rev. for the Sun in a year	–	•	•	–	–	•
92	Excess of revolution for the Sun in collected years	–	•	•	–	–	•
93	Excess of rev. for the Sun due to solar anomaly	–	•	•	–	–	•
94	Excess of rev. for the Moon and the ascending node in collected years	–	•	•	–	–	•
95	Excess of rev. for the Moon and the ascending node in expanded years	–	•	•	–	–	•
96	Mean motion of the lunar node in hours	–	•	•	–	–	•
97	Excess of rev. for Saturn in collected years	–	•	•	–	–	•
98	Excess of rev. for Jupiter in collected years	–	•	•	–	–	•
99	Excess of rev. for Mars in collected years	–	•	•	–	–	•
100	Excess of rev. for Venus in collected years	–	•	•	–	–	•
101	Excess of rev. for Mercury in collected years	–	•	•	–	–	•
102	Excess of rev. for Saturn in expanded years	–	•	•	–	–	•
103	Excess of rev. for Jupiter in expanded years	–	•	•	–	–	•
104	Excess of rev. for Mars in expanded years	–	•	•	–	–	•
105	Excess of rev. for Venus in expanded years	–	•	•	–	–	•
106	Excess of rev. for Mercury in expanded years	–	•	•	–	–	•
107	Animodar	–	–	–	–	–	•
108	Mean motion of the Moon in days	–	–	–	–	–	•
109	Mean motion of the Moon in hours	–	–	–	–	–	•
110	Mean motion of the Sun in hours	–	–	–	–	–	•
111	Mean motion of the Sun in minutes of an hour	–	–	–	–	–	•
112	Conversion of degrees into hours	–	–	–	–	–	•

We have followed this numbering when reproducing Bianchini's tables, whether fully or partially, in the text below. In some cases we have added a letter after these numbers, to refer to a table compiled by us; or to a table by Bianchini, presented differently; or to a table taken from another source.

2. TABLES IN ED. 1495

Table 1. Precession

Na, ff. 14r–15r: *Tabula augis communis et veri motus octave spere*
Nu, f. 21r: *Tabula veri motus octave spere et augium planetarum*
Va, ff. 185r–186v: *Tabula augis communis et veri motus octave spere*
Ed. 1495, ff. a2r–a6v: *Tabula motus augium communium*
Ed. 1526, ff. 2r–6v: *Tabula motus augium communium*

As in the Parisian Alfonsine Tables, precession or, more properly, trepidation (i.e., 'access and recess' as it was called by many medieval astronomers), is here applied to the longitudes of fixed stars given in Ptolemy's star catalogue and to the planetary apogees. The usual distinction is that precession refers to a uniform motion of the 8th sphere (the sphere of the fixed stars) whereas trepidation refers to a variable motion of that sphere. But here we use the term 'precession' to refer to any motion ascribed to the 8th sphere. The epoch of this table is the date taken by Bianchini as that of Ptolemy's star catalogue: May 17, 16 A.D. (see p. 32, below).

Table 1 has 3 columns. The first is for the number of collected years, generally given at intervals of 60 years, from 60 to 49000. However, the argument is given at intervals of 10 years when the magnitude—which we shall call p_2—is near its extremal values (at arguments 1750y, 5250y, and so on, with a periodicity of 3500y; see Table 1A, below). The second column, headed *locus augium in 9 spera* (Na), or simply *locus augium*, displays the increment to the apogees, that is, the precession (in physical signs of 60°, degrees, minutes, and seconds), p, with respect to their values at epoch. Let the apogee of a given planet at time t_1 be A_1, and its apogee at time t_2 be A_2, where $t_2 > t_1$; then

$$p = A_2 - A_1.$$

Table 1: Precession (excerpt)

Time (years)	Locus Augium (°)	Motus in anno (°)
	Direct	Add
60	0;55,25	0; 0,55,14
120	1;50,39	0; 0,55, 5
180	2;45,44	0; 0,55, 3
...		
2960	25;55,32	0; 0, 0,31
	Retrograde	Subtract
3020	25;56, 3	0; 0, 0,24
...		
3920	25;29,58	0; 0, 0,14
	Direct	Add
3980	25;29,49	0; 0, 0,31
...		
7000	51;25,45	0; 0,55,14
7060	52;21	–
...		
48940	5,59; 0	–
49000	0; 0	–

Although p is always positive when A_1 corresponds to the epoch of this table, this is not always the case for an arbitrary time t_1. Hence entries in the table from the radix (year 0) to year 2960 increase (here called 'direct'), but from year 3020 to year 3980 they decrease (here called 'retrograde').

The third column, headed *motus in anno*, gives the yearly increment or decrement (in seconds and thirds) for purposes of interpolation, and its entries can be recomputed by dividing the difference between two consecutive entries in the preceding column by the number of years between their arguments. In both printed editions the entries for this third column are only given for the interval (60y, 7000y).

The entries for the increments in the position of the apogees result from adding two components, a linear term and a periodic term (see Figure 1).

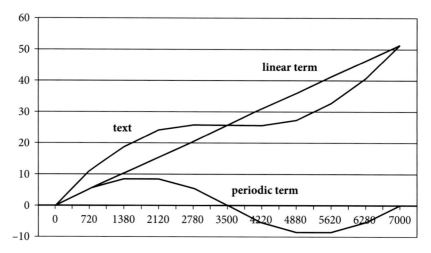

Figure 1: Components of precession for 7,000 years, beginning with
the radix.

The linear component is given by the expression

$$p_1 = 360 \, t \, / \, 49000,$$

whereas the periodic component agrees very closely with the function

$$p_2 = \arcsin (\sin 9° \cdot \sin (360 \, t \, / \, 7000)),$$

where t, the number of years, is the argument in the table. Both periods
(7000 years and 49000 years) and the amplitude (9°) are those typically
used in the Parisian Alfonsine Tables. Thus, the total precession can be
written as

$$p = p_1 + p_2$$

or

$$p = p_1 + \arcsin (\sin 9° \cdot \sin 7 \, p_1).$$

Table 1A displays an excerpt of the column for the increment of the
apogees (under the heading 'Text'), as well as the results of our computations for p_1 (labeled 'linear term' in Fig. 1), p_2 (labeled 'periodic term'
in Fig. 1), and p.

Table 1A: Recomputation of precession

Time (years)	Text (°)	p_1 (°)	p_2 (°)	p (°)
	Direct			
720	10;41,43	5;17,23	5;24,20	10;41,43
...				
1380	18;38,37	10; 8,20	8;30,16	18;38,36
...				
1750	21;51,26	12;51,26	9; 0, 0	21;51,26
...				
2120	24; 4,50	15;34,32	8;30,16	24; 4,48
...				
2780	25;49,49	20;25,28	5;24,20	25;49,48
...				
2960	25;55,32	21;44,49	4;10,48	25;55,37
	Retrograde			
3020	25;56, 3	22;11,16	3;44,46	25;56, 2
...				
3500	25;42,53	25;42,51	0; 0, 0	25;42,51
...				
3920	25;29,58	28;48, 0	–3;18, 5	25;29,55
	Direct			
3980	25;29,49	29;14,27	–3;44,46	25;29,51
...				
4220	25;35,58	31; 0,15	–5;24,20	25;35,55
...				
4880	27;20,57	35;55,12	–8;30,16	27;20,56
...				
5250	29;34,20	38;34,17	–9; 0, 0	29;34,17
...				
5620	32;47, 7	41;17,23	–8;30,16	32;47, 7
...				
6280	40;44, 2	46; 8,20	–5;24,20	40;44, 0
...				
7000	51;25,45	51;25,43	0; 0, 0	51;25,43

The standard Parisian Alfonsine Tables, as presented in the *editio princeps* (1483), require the use of two tables, one for the mean motion of access and recess of the 8th sphere, and another for its equation.[1] In contrast, Bianchini has merged these two tables into one.

---

[1] Ratdolt, *Tabule astronomice Alfontij*, d3v–d4r; see also Poulle, *Les tables alphonsines*, 131–32.

Note that the expression used here for p_2 differs from the one usually, but erroneously, attributed to Alfonsine astronomy, $p_2 = 9° \cdot \sin (360 t / 7000)$.[2] This misconception was based on the assumption that the formula underlying this periodic component (namely the table for the equation of access and recess on the 8th sphere) was a sinusoidal function of the form

$$y = 9° \cdot \sin x, \qquad [1]$$

where x is the mean motion of access and recess. However, it is not; rather, it corresponds to a function of the form

$$y = \arcsin (\sin 9° \cdot \sin x), \qquad [2]$$

which is equivalent to

$$\sin y = \sin 9° \cdot \sin x. \qquad [3]$$

This is readily verified in Table 1B where we display an excerpt of the entries in the Alfonsine table for the equation of access and recess in the eighth sphere and compare them with the values resulting from formulas [1] and [2].

It is thus clear that formula [2], related to the solution of a spherical right triangle with an arc of 9° as hypotenuse and y as the side opposite angle x, reproduces the tabulated entries, in contrast to equation [1], related to the solution of a plane right triangle. And there is also no doubt that Bianchini strictly followed the Alfonsine tradition in this respect.

The use of this single table is explained in Chapter 3 (ed. 1526). We are told that the number of years serving as argument in this table has to be counted from an initial date, 15 years and 137 days from the beginning of the common era. Regiomontanus reproduced this information at the bottom of f. 21r in his copy of Bianchini's tables. This date corresponds to May 17, 16 A.D. This is also a characteristic feature of the Parisian Alfonsine Tables as has been demonstrated about 30 years ago for the first time in the modern literature.[3] Indeed, Bianchini's canons are the earliest source we have found that gives this date explicitly for the epoch of the Alfonsine model. This means that Bianchini accepted 16 A.D. as the date of Ptolemy's star catalogue even though in the *Almagest*

² See, e.g., Chabás and Goldstein, *Alfonsine Tables of Toledo*, 260–61.
³ Mercier, "Medieval Conception of Precession," 58–60.

Table 1B: Recomputation of the Alfonsine table for the equation of access
and recess in the eighth sphere*

Mean Mot. acc. & rec. (°)	Equation in Alf. T. 1483 (°)	Equation computed from [1] (°)	Equation computed from [2] (°)
10	1;33,20	1;33, 6	1;33,24
20	3; 3,49**	3; 4,41	3; 4, 1
30	4;29,10	4;30, 0	4;29,10
40	5;46,16	5;47, 6	5;46,16
50	6;52,58	6;53,40	6;52,57
60	7;47,10	7;47,39	7;47,10
70	8;27,11	8;27,26	8;27,11
80	8;51,44	8;51,48	8;51,44
90	9; 0, 0	9; 0, 0	9; 0, 0

* Ratdolt, *Tabule astronomice Alfontij*, d3v.
** Most of the manuscripts we have consulted have this entry, but in two of them
the tabulated value is closer to what we recomputed from formula [2]: 3;3,59° (Paris,
Bibliothèque nationale de France, 7286C, f. 10r, a manuscript associated with John
Vimond and John of Lignères), and 3;4,8° (Bologna, University Library, MS 2248,
f. 17v, a manuscript associated with Prosdocimo de' Beldomandi; see Chabás 2007).
The entry in the 2nd edition of the Alfonsine Tables (1492) is the same as in the *editio
princeps* (1483).

Ptolemy gives a date equivalent to 137 A.D. The medieval traditions for
this date range from 134 to 140.[4] We believe that the importance given
to this date by Bianchini as a zero point for trepidation is the result of
his constant focus on cyclical epochs, as we find in his tables for the
Sun, the Moon, and the planets.

Chapter 3 offers two examples for using the table: the first is for June
13, 1446 at 8;50h, and the second for a date long before the beginning
of the common era: 3429 years and 164 days before it. According to
the text, the difference between the proposed date and the radix in the
first example is 1430 years and 27 days. This date serves as argument
in the table, where we find 18;38,37° for 1380 years, and 0;0,35,14° for
the corresponding yearly increment. Multiplying the increment by 50
(= 1430 − 1380) and adding the result to the entry for 1380 years (as
indicated in the text), we obtain 19;7,58,40°, in contrast to the value
19;7,50,40° (≈ 19;8°) found in Bianchini's text. Strictly speaking, this

---

[4] North, "Just whose were the Alfonsine Tables?," 464–65.

approximate value corresponds to the end of year 1445, and it does not take into account fractions of a year.

Table 2. Radices of the apogees

Na, f. 15r: *Tab* (...) *pla* (...)
Nu, f. 21r: *Radices augium*
Va, f. 15r: *Tabula radice augium planetarum*
Ed. 1495, f. a6v: *Radix augium*
Ed. 1526, f. 6v: *Radix augium*

This table lists the radices of the apogees, reproduced in Table 2A, where we have added a column for the corresponding entries for the Incarnation in the ed. of 1483 of the Alfonsine Tables.

Table 2A: Radices of the apogees

	Bianchini (°)	Incarnation Alf. T. 1483 (°)
Sun and Venus	1,11;32, 8	1,11;25,23
Saturn	3,53;30,27	3,53;23,42
Jupiter	2,33;43,45*	2,33;37, 0
Mars	1,55;18,58	1,55;12,13
Mercury	3,10;46,19	3,10;39,33

* Both printed editions have 2,33;43,<u>55</u>°.

The apogees listed by Bianchini correspond to year 16 A.D., taking precession into account, and they are consistent with the Alfonsine values shown in Table 2A for year 1 A.D. which do not take precession into account. For example, computing the solar apogee with the standard version of the Parisian Alfonsine Tables one finds 1,11;32,8° for May 17, 16 A.D. (in agreement with Bianchini) and 1,11;17,56° for the Incarnation. This implies that precession amounted to −0;7,27° (= 1,11;17,56° − 1,11;25,23°) at the Incarnation, a quantity given as [−]0;7,25° in ed. 1483 of the Alfonsine Tables.[5] Similar agreement is found for the rest of the planets. In short, for the apogees Bianchini faithfully followed the Alfonsine tradition.

---

[5] Ratdolt, *Tabule astronomice Alfontij*, c8r.

Chapter 4 gives two examples for using the table for the same dates as in Chapter 3 (see Table 1 for precession): the first is for June 13, 1446 at 8;50h and the second for a date long before the beginning of the common era. According to the text, to obtain the position of the solar apogee on the given date, we have to add the increment of precession since the epoch found previously for the first date (19;8°) to the radix for the apogee of the Sun and Venus (1,11;32°): the result is 1,30;40° (or 90;40°). As was the case for precession, strictly speaking this value corresponds to the end of year 1445. In Chapter 6, which contains a continuation of this worked example, we are told that a more precise value for the solar apogee at that time is 1,30;40,17°, and this is a reasonable value for the solar apogee for this date (see also Table 8).

Table 3. Days in a year beginning in January

Na, f. 15v: *Tabula solis et capitis draconis*
Nu, f. 31v: *Menses aliorum quatuor*
Va, f. 186v: *Tabula dierum anni a Januario et alia a Februario*
Ed. 1495, f. a6v: *Tabula dierum mensium*
Ed. 1526, f. 6v: *Tabula dierum mensium*

The text displays the accumulated numbers of days at the end of each month from January to December in two columns, one for a common year and the other for a leap year: see Table 3. For an analogous table beginning in June, see Table 10 below.

Table 3: Days in a year beginning in January (excerpt)

Month	Common y. (d)	Leap y. (d)
January	31	31
February	59	60
...		
November	334	335
December	365	366

Chapter 2 gives an example for using this table for July 17, 1447, that is, 198 days after the beginning of the year.

Table 4. Mean motion of the Sun and the lunar node in years

Na, f. 15v: *Tabula solis et capitis draconis*
Nu, f. 30r: *Tabula veri motus capitis draconis*
Va, f. 186v: *Tabula motus* [symbol for the Sun] *et* [symbol for the
  ascending node] *draconis*
Ed. 1495, f. a7r: *Tabula solis et capitis draconis in medijs motibus*
Ed. 1526, f. 7r: *Tabula* [symbol for the Sun] *et* [symbol for the ascend-
  ing node] *draconis in mediis motibus*

This table has three columns: see Table 4. The first displays the argument
(in years, in descending order, from 1000y to 100y at intervals of 100y,
from 100y to 20y at intervals of 10y, and from 20y to 1y at intervals of
1y). The second column displays the mean motion of the Sun (in physical
signs, degrees, minutes, and seconds), and the third displays the mean
motion of the lunar ascending node (in physical signs, degrees, minutes,
and seconds). The entries for 1000y correspond to daily mean motions
of 0;59,8,19,37,19°/d and –0;3,10,38,7,14°/d, respectively. As expected,
these are the standard parameters used in the Alfonsine corpus.

Regiomontanus's copy has no column for the solar mean motion, and
the entries for the mean motion of the lunar node are the complement
in 360° of those in Table 4.

Table 4: Mean motion of the Sun and the lunar node in years (excerpt)

Years	Sun (°)	Lunar node (°)
1000	7;20,48	4,21;32,39
900	6;36,44	2, 7;23,23
...		
2	5,59;31,20	38;39,24
1	5,59;45,39	19;19,42

Table 5. Radices of the Sun and the lunar node

Na, f. 15v: *Tabula solis et capitis draconis*
Nu, f. 30r: *Tabula veri motus capitis draconis*
Va, f. 186v: *Tabula radicis domini nostri Iesu Cristi incarnati*
Ed. 1495, f. a7r: *Tabula radicum*
Ed. 1526, f. 7r: *Tabula radicum*

This table has the same layout as that of Table 4, and gives the radices for the Incarnation, 1400, and 1440: see Table 5. Ed. 1526 adds entries for 1480 and 1500. As was the case with Table 4, Regiomontanus only listed the mean motion of the lunar node and presented it as the complement in 360° to the entries in Table 5. However, Regiomontanus has 4,58;4,18° for the Incarnation and 3,39;54,34° for 1400, whereas he should have 4,28;4,18° and 3,9;54,34°, respectively.

Table 5: Radices of the mean Sun and the lunar node

	Sun (°)	Lunar node (°)
Incarnation	4,38;17,34	1,31;55,42
1400	4,48;34,42	2,50; 5,27
1440	4,48;52,20	3,43;45, 9

Both entries for the Incarnation follow from the standard Alfonsine table for the radices for places other than Toledo for a geographical longitude of 32°.[6] This is exactly the longitude of Ferrara given in Bianchini's text (Introduction and Chapter 1), but differs from that assigned to Ferrara (33;30°) in the table for geographical coordinates, which is only included in ed. 1526 (see Table 87, below). The difference in longitude between Toledo and Ferrara is 21°, and corresponds to adjustments of −0;3,27° and +0;0,11° in the values for the mean motion of the Sun and the lunar node, respectively.

The entries for the radices in 1400 and 1440 in Table 5 are obtained by adding those for the Incarnation (computed for a geographical longitude of 32°) to those corresponding to 1400 and 1440 years in the table of the mean motions. Thus, for 1400 we find for the Sun and the lunar node, respectively:

$$4,48;34,42° = 4,38;17,34° \text{ (Incarn.)} + 7;20,48° \text{ (1000y)} + 2;56,20° \text{ (400y)},$$

and

$$2,50;5,25° = 1,31;55,42° \text{ (Incarn.)} + 4;21,32,39° \text{ (1000y)} + 2;56,37,4°$$
$$\text{(400y)},$$

---

[6] Ratdolt, *Tabule astronomice Alfontij*, d1v–d2v; see also Poulle, *Les tables alphonsines*, 127–29.

in agreement with the tabulated values. We note, however, that these values correspond to the positions of the Sun and the lunar node for 1400 complete years after the Incarnation, not at the beginning of year 1400. Thus the 'radices for 1400', as expressed by Bianchini, have to be understood as the values at the beginning of year 1401 for a place at geographical longitude 32°.

Table 6. Mean motion of the Sun and the lunar node in days

Na, f. 15v: *Tabula solis et capitis draconis*
Nu, f. 30r: *Tabula veri motus capitis draconis*
Va, f. 186v: *Tabula motus* [symbol for the Sun] *et* [symbol for the ascending node] *draconis*
Ed. 1495, f. a7r: *Medii motus in diebus*
Ed. 1526, f. 7r: *Medii motus in diebus*

This table, with the same layout as that of Table 4, gives the mean motion of the Sun and the lunar node for a number of days (in decreasing order, from 300d to 100d at intervals of 100d, from 100d to 10d at intervals of 10d, and from 10d to 1d at intervals of 1d): see Table 6. In MS Nu only the mean motions of the lunar node from 1d to 31d, at intervals of 1d, are given.

Table 6: Mean motion of the Sun and the lunar node in days (excerpt)

Days	Sun (°)	Lunar node (°)
300	4,55;41,37	15;53,11
200	3,17; 7,45	10;35,27
...		
2	1;58,17	0; 6,21
1	0;59, 8	0; 3,11

Table 7. Mean motion of the Sun and the lunar node in hours

Na, f. 15v: *Tabula solis et capitis draconis*
Nu, f. 30r: *Tabula veri motus capitis draconis*
Va, f. 186v: *Tabula motus* [symbol for the Sun] *et* [symbol for the ascending node] *draconis*
Ed. 1495, f. a7r: *Tabula medii motus in horis*
Ed. 1526, f. 7v: *Tabula medii* [symbol for the Sun] *et* [symbol for the ascending node] *motus in horis*

This table, with the same layout as that of Table 4, gives the mean motion of the Sun and the lunar node for a number of hours (from 1h to 30h at intervals of 1h, and from 30h to 60h at intervals of 10h): see Table 7. In Regiomontanus's copy, we are only given the mean motion of the lunar node, as in the previous tables.

Table 7: Mean motion of the Sun and the lunar node in hours (excerpt)

Hours	Sun (°)	Lunar node (°)
1	0; 2,28	0; 0, 8
2	0; 4,56	0; 0,16
...		
50	2; 3,10	0; 6,38
60	2;27,48	0; 7,58

Chapter 5 gives an example for using the tables for the mean motion of the Sun for May 12, 1447 at 15;20,30h. The result, after selecting the entries in the various tables, is

$$2,20;52,1° = 7;20,48° \ (1000y) + 2;56,20° \ (400y) + 0;17,38° \ (40y) + 5,59;33,9° \ (6y) + 1,38;33,52 \ (100d) + 29;34,10° \ (30d) + 1;58,17° \ (2d) + 0;36,58° \ (15h) + 0;0,49° \ (20 \ min) + 0;0,0° \ (30s).$$

The mean longitude of the Sun, 59;9,35°, is obtained by adding the previous value to the radix of the Incarnation (4,38;17,34°).

There were two conventions in the Middle Ages for dates in the Julian calendar. One convention was to use complete years, months, and days, and the other was to use current years, months, and days. The worked example presented in Chapter 5 for May 12, 1447 indicates that Bianchini employs a mixed system: the given date uses a current year and a current month, but complete days (e.g., 'May 12' has to be understood as '132 days have elapsed since the beginning of the year' or '12 days of May have elapsed'). The epoch for this calendar is the Nativity which, in this context, means noon of the day preceding Jan. 1, 1 A.D. In other words, at the given date, only 1446 and 4 months have elapsed. This is not the criterion used nowadays, that is, May 12, 1447 means that 1447 is the current year (1446 complete years have elapsed since epoch), May is the current month (4 complete months have elapsed in the current year), and day 12 means that 11 complete days have elapsed in the current month. Moreover, the Parisian Alfonsine Tables and the *editio princeps* of the Alfonsine Tables use current years and months and, for both, 'May 12' means that '11 days of May have elapsed'. Ultimately, Bianchini's criterion might derive from the way he constructed his tables, for they are based on adding excesses of days for different dates in order to find the correct time for which a quantity is to be computed. With this criterion, it would seem easier then to work with complete rather than incomplete days.

Table 8. Position of the solar apogee

Na, ff. 16r–17v: *Tabula solis in auge*
Nu, f. 21v: *Tabula ingressus solis in augem*
Va, ff. 187v–189r: *Tabula ad inveniendum introitum solis in augem*
Ed. 1495, ff. a8r–b3v: *Tabula solis in auge*
Ed. 1526, ff. 8r–11v: *Tabula solis* [symbol for the Sun] *in auge*

In this table there are five columns: see Table 8. The first is for the argument: collected years from 0y to 2000y at intervals of 4y (at intervals of 20y in Regiomontanus's copy). Thus, all the entries correspond to leap years. The entries in the second column, under the heading 'June', are given in days, hours, and minutes (from 3d 23;38h for year 0 to 13d 1;6h for year 2000). As explained in Chapter 7, the entries show the date and the time the Sun is at its apogee in a given year. The heading of the third column is 'equation' and entries are only given for some selected arguments, in hours, minutes, and seconds; they represent the amount to be added to the corresponding entry in the second column when the given year is one, two, or three years after a leap year. Column 4 displays the longitude of the solar apogee, in physical signs, degrees, minutes, and seconds, whereas column 5, headed *motus in anno*, represents the yearly progress (in seconds) of the solar apogee. We note that the entries in column 5 are the same, although with a lower precision, as the corresponding ones in Table 1 for the yearly motion of the apogees.

The instructions for using this table are found in Chapter 7 where the worked example consists in determining for year 1446 the position of the solar apogee and the time when the Sun reaches it. The date and time sought are June 13 and 8;50h, and this date was used previously in the discussion of Tables 1 and 2. The resulting position of the solar apogee is said to be 1,30;40,16°. This value appears as 1,30;40,17° in Chapter 6 (see our comments to Table 2).

Table 8: Position of the solar apogee (excerpt)

Coll. Years	June (d, h)		Equation (h)	Apogee (°)	Motion (s)
0	3	23;38	1 after leap y.	1,11;17,32	55
4	4	0;25	6;11,46	1,11;21,14	55
8	4	1;22		1,11;24,56	55
12	4	1;59		1,11;28,38	55
16	4	2;46		1,11;32,19	55
...					
1444	12	20;44	2 after leap y.	1,30;39, 8	34
1448	12	20;57	12; 6,26	1,30;41,25	34
...					
2000	13	1; 6	17;53, 9	1,34;54,48	20

This table also shows that the radix for the solar apogee given in Table 2 (1,11;32,8°) corresponds to a little less than 16 complete years.

Table 9. Planetary apogees

Na, f. 17v: *Adde augi ad habendum auges aliorum planetarum*
Nu, f. 21r: *Ad habendum auges aliorum adde augi solis*
Va, f. 189r: *Adde augi solis ad habendum auges aliorum planetarum*
Ed. 1495, f. b3v: *Adde augi* [symbol for the Sun] *ad habendum auges aliorum planetarum*
Ed. 1526, f. 11v: *Adde augi* [symbol for the Sun] *ad habendum auges aliorum planetarum*

As indicated in the title, this table displays the differences between the planetary apogees of the planets and the solar apogee: see Table 9. The apogees of the planets and the Sun are listed in Table 2 and we reproduced them in Table 2A, above.

Table 9: The differences between the planetary apogees and the solar apogee

	Diff. (°)
Saturn	2,41;58,19
Jupiter	1,22;11,37
Mars	43;46,50
Venus	0; 0, 0
Mercury	1,59;14,11

Chapter 8 (ed. 1526) reproduces the entries in Table 9 with no further explanation, but we observe that they were obtained by subtracting the radix of the solar apogee for year 16 A.D. (1,11;32,8°) from each of the radices of the planetary apogees for the same date listed in Table 2A. Thus, for Saturn, 2,41;58,19° = 3,53;30,27° – 1,11;32,8°, and similarly for the rest of the planets. And this is indeed the explicit instruction given in the planetary tables, below, where we read, *Pro habenda auge* [symbol for Saturn] *adde auge* [symbol for the Sun] 2,41;58,19°, in the case of Saturn.[7] For the rest of the planets, similar instructions are given, but not for Venus.

Table 10. Days in a year beginning in June

Na, f. 17v: *Menses non bisextiles* and *Menses bisextiles*
Va, f. 186v: *Tabula dierum anni a Januario et alia a Februario*
Ed. 1495, f. b3v: *Menses non bisextiles* and *Menses bisextiles*
Ed. 1526, f. 11v: *Menses non bisextiles* and *Menses bisextiles*

This table displays the accumulated numbers of days at the end of each month from June to May in two columns, one for a common year and the other for a leap year: see Table 10. For an analogous table beginning in January, see Table 3 above.

Table 10: Days in a year beginning in June (excerpt)

Month	Common y. (d)	Leap y. (d)
June	30	30
July	51	61
...		
April	334	335
May	365	366

---

[7] Bianchini, *Tabule Blanchini* (1526), 27v.

Table 11. True motion of the Sun

Na, f. 18r–v: *Tabula veri motus solis ab auge*
Nu, ff. 22v–23r: *Tabula veri motus solis ab auge*
Va, ff. 189v–190r: *Tabula veri motus solis ab auge*
Ed. 1495, ff. b4r–b5v: *Tabula veri motus solis ab auge*
Ed. 1526, ff. 12r–13v: *Tabula veri motus* [symbol for the Sun] *solis ab auge*

This table has three columns: see Table 11. The first is for the argument, the number of days (from 0d to 365d) since the date the Sun reached apogee in a given year. The second displays the true motion of the Sun (in physical signs, degrees, minutes, and seconds) corresponding to this time, and the third gives the hourly solar velocity in minutes and seconds (from 0;2,23°/h at 0d and 365d to a maximum of 0;2,34°/h at about 182d).

Table 11: True motion of the Sun (excerpt)

Days	True motion (°)	Velocity (°/h)
0	0; 0, 0	0;2,23
1	0;57, 1	0;2,23
2	1;54, 2	0;2,23
...		
181	2,58;20,17	0;2,34
182	2,59;21,48*	0;2,34
183	3, 0;23,19	0;2,34
184	3, 1;24,49	0;2,34
...		
363	5,57;52, 9	0;2,23
364	5,58;49,10	0;2,23
365	5,59;46,11	0;2,23

* Ed. 1526 has 2,59;21,45°.

We note that the minimum progress of the true motion (0;57,1°) occurs between days 0 and 1 and between days 364 and 365, whereas the maximum progress (1;1,34°) occurs between days 182 and 183. These values for daily progress correspond to 0;2,22,55°/h and 0;2,23,55°/h, respectively, which seem to have been rounded by the author to 0;2,23°/h and 0;2,34°/h.

Regiomontanus chose to use signs of 30° in this particular case, rather than signs of 60° used elsewhere, an indication that he decided not to reproduce the table computed by Bianchini. Moreover, Regiomontanus's entries are systematically slightly different from those of Bianchini, and an extra line was added for 366d, as shown in Table 11A.

Table 11A: True motion of the Sun (Regiomontanus)

Days	True motion (s, °)		Velocity (°/h)
0	0;	0, 0	0;2,22
1		0;56,59	0;2,22
2		1;53,58	0;2,22
...			
181	5	28;20,22	0;2,34
182	5	29;21,50	0;2,34
183	6	0;23,18	0;2,34
184	6	1;24,46	0;2,34
...			
363	11	27;52,13	0;2,22
364	11	28;49,12	0;2,22
365	11	29;46,11	0;2,22
366	0	0;43,10	0;2,22

To determine the maximum solar equation embedded in this table, consider first the daily mean velocity of the Sun derived from the entry for 365 days: 5,59;46,11°/365d = 0;59,8,24,49,18°/d. By inspecting the table, one sees that 0;59,8° is the difference between days 93 (1,29;29,55°) and 94 (1,30;29,3°), and between days 94 (1,30;29,3°) and 95 (1,31;28,11°). Now, computing the position of the mean Sun for days 93, 94, and 95, we obtain 1,31;40,2°, 1,32;39,10°, and 1,33;38,19°, respectively. If we subtract these mean positions from the true positions specified in the table, we obtain 2;10,7°, 2;10,7°, and 2;10,8°, respectively. These are the maximum differences between mean and true positions of the Sun, that is, the maximum solar equation, which we can take confidently as 2;10°. This is indeed the value used in the standard Alfonsine Tables.

As is the case with other tables, Bianchini's text offers an example of its use (Chapter 9). We are asked to find the true position of the Sun in 1446, August 16 at 15;40h. As explained in Chapter 7 (see comments to Table 8, above), in 1446 the Sun reaches its apogee on June 13 at 8;50h and the position of the solar apogee is 1,30;40,16°. The time since apogee is 64d 6;50h, and therefore 64d is taken as the argument for Table 11. The corresponding entries for 64d are 1,1;10,48° (true motion) and 0;2,25°/h (hourly motion), and the true motion of the Sun that was sought is

$$2,32;7,36° = 1,30;40,16° + 1,1;10,48° + (6;50\text{h} \cdot 0;2,25°/\text{h}).$$

This is the value given in the text, but we note that ed. 1526 has 2,32;7,34°.

Table 12. Mean motion of the Moon in collected years

Na, f. 19r: *Tabula radicum lune*
Nu, f. 23v: *Tabula radicum lune*
Va, f. 190v: *Tabula radicum lune*
Ed. 1495, f. b6r–v: *Tabula radicum lune in annis collectis*
Ed. 1526, f. 14r–v: *Tabula radicum* [symbol for the Moon] *in annis collectis*

In this table there are five columns: see Table 12. The first is for the argument: collected years from 40 to 2400 at intervals of 40y (only up to 2000y in Regiomontanus's copy). Thus, all the entries correspond to leap years. The entries in the other columns represent the time (in days, hours, and minutes), the double elongation (in degrees and minutes), the longitude of the Moon (in physical signs, degrees, and minutes), and the argument of lunar latitude (in physical signs, degrees, and minutes). In Regiomontanus's copy the column for the argument of lunar latitude was omitted. The time displayed in column 2 is the excess over an integer number of anomalistic months; hence, the entries in this column do not exceed 27d 13;18,36h, which is the length of the anomalistic month mentioned in Chapter 11 and explicitly given in Table 14, below. We note that the author uses degrees from 0 to 360 for the double elongation, here called *centrum* and, on the other hand, physical signs of 60° for the lunar longitude and the argument of lunar latitude.

Table 12: Mean motion of the Moon in collected years (excerpt)

Years	Time (d, h)		Double elongation (°)	Longitude (°)	Arg. latitude (°)
40	6	2; 0	25;20	3, 6;58	4, 0;19
80	12	4; 1	50;41	13;56	2, 0;37
...					
2360	0	17;15	148;42	4,30;58	3,17; 1
2400	6	19;15	174; 3	1,37;56	1,17;18

Table 13. Mean motion of the Moon in expanded years

Na, f. 19r: *Tabula radicum lune*
Nu, f. 23v: *Tabula radicum lune*
Va, f. 190v: *Tabula radicum lune*
Ed. 1495, f. b7r–v: *Tabula radicum lune in annis expansis*
Ed. 1526, f. 15r–v: *Tabula radicum lune in annis expansis*

The layout of this table is the same as that of Table 12, but here the argument is the number of expanded years, from 1y to 40y: see Table 13. The entries for 40 expanded years agree with those for 40 collected years in Table 12. As was the case in the previous table, Regiomontanus omitted the column for the argument of lunar latitude.

Table 13: Mean motion of the Moon in expanded years (excerpt)

Years	Time (d, h)		Double elongation (°)	Longitude (°)	Arg. latitude (°)
1	6	18;59	93;41	39;54	58;52
2	13	13;57	187;21	1,19;49	1,57;45
...					
39	25	20;20	339;50	2,24; 0	2,56;55
40	6	2; 0	25;20	3, 6;58	4, 0;19

In order to understand the peculiar organization of Bianchini's tables, it is essential to realize that the entries in Tables 12 and 13 for the double elongation, the lunar longitude, and the argument of lunar latitude, do not correspond to the exact number of years indicated in the first column, but to the closest previous time when the lunar anomaly was 0°.

For instance, in the case of the entry for 1 year, 6d 18;59h means that in a year of 365 days there is an integer number of anomalistic months plus 6d 18;59h and, indeed, 6d 18;59h = 365d – (13 · 27d 13;18,36h). Thus, the entry for 1 year of 93;41° for the double elongation has to be understood as the increment in double elongation between the beginning of the year and this precise moment (6d 18;59h before the completion of the year) at the rate of 2 · 12;11,27°/d, the standard parameter used in Alfonsine astronomy for the double elongation.

Table 14. Multiples of the mean anomalistic month

Na, f. 19r: *Tabula brevis*
Nu, f. 23v: *Brevis tabula*
Va, f. 190v: *Tabula brevis*
Ed. 1495, f. b7v: *Tabula brevis*
Ed. 1526, f. 15v: *Tabula brevis*

The five columns in this table are the same as in Table 12, but in this case we are only given entries for the first four multiples of the mean anomalistic month: see Table 14. The purpose of this table is to facilitate computation, and it is used when the accumulated entries for collected years, expanded years, months, days, etc. exceed one or more anomalistic months, which then have to be subtracted (thus, the heading *minue*); see the worked example in our comments to Table 17, below. This is one of many instances where Bianchini pays special attention to finding the time when the mean anomaly is 0°. As was the case in the Tables 12 and 13, Regiomontanus omitted the column for the argument of lunar latitude.

Table 14: Multiples of the mean anomalistic month

Days	*Minue* (h)	Double elongation (°)	Longitude (°)	Arg. latitude (°)
27	13;18,36	311;49,19	3; 4,12	4;31,45
55	2;37,12	263;39, 0	6; 8	9; 3
82	15;55,48	215;28, 0	9;13	13;35
110	5;14,24	167;17, 0	12;17	18; 7

The entries for 27d 13;18,36h, the length of the anomalistic month, are also derived from the standard parameters in Alfonsine astronomy, for

$$311;49,19° = 2 \cdot (27d\ 13;18,36h \cdot 12;11,26,41°/d) - 360°,$$
$$3;4,12° = (27d\ 13;18,36h \cdot 13;10,35,1°/d) - 360°,$$
$$4;31,45° = (27d\ 13;18,36h \cdot 13;13,45,39°/d) - 360°,$$

where 12;11,26,41°/d, 13;10,35,1°/d, and 13;13,45,39°/d are, respectively, the mean motions in elongation, lunar longitude, and argument of lunar latitude, used in the Parisian Alfonsine Tables.

Table 15. Radices of the Moon in anomaly

Na, f. 19r: *Tabula radicis*
Nu, f. 23v: *Tabella radicum*
Va, f. 190v: *Tabula radicis*
Ed. 1495, f. b7v: *Tabula radicum*
Ed. 1526, f. 15v: *Tabula radicum*

This table for the radices of the Moon when its anomaly is 0° has four columns: see Table 15.

Table 15: Radices of the Moon

		Time	Double elongation	Longitude	Arg. latitude
		(d, h)	(°)	(°)	(°)
Incarnation	15	4;10	37;29	4,42; 5	13;12
1400	7	15;51	178;57	10;30	3, 0;11
1440	13	17;51	204;17	3,17;28	1, 0;30

As noted previously, Regiomontanus omitted the column for the argument of lunar latitude, but he added the radix for Vienna (Incarnation): 15d 4,14h. In ed. 1526 three rows were added for years 1480, 1500, and 1508.

The entry 15d 4;10h for the Incarnation has to be understood as meaning that 15d 4;10h before the Incarnation the mean lunar anomaly was 0°. According to the Parisian Alfonsine Tables, the lunar anomaly for the Incarnation and the longitude of Ferrara was 3,18;14,31°.[8] The time taken by the Moon to cover this distance, at the mean motion of anomaly used in Alfonsine astronomy (13;3,54°/d) is precisely 15d 4;10h (= 3,18;14,31/13;3,54), in agreement with the entry given in Table 15.

The entries for the double elongation, the lunar longitude, and the argument of lunar latitude given for the Incarnation are the values of these quantities at the time when the anomaly was 0° (that is, Incarnation – 15d 4;10h), and can also be computed from the tables of radices in the Parisian Alfonsine Tables. Thus, for double elongation, at the Incarnation for the geographical longitude of Ferrara, the longitude of the Sun was 4,38;17,34° (see Table 5) and that of the Moon 2,2;0,43° = 122;0,43°;[9] hence, the double elongation was 407;26,18° = 47;26,18°. Now, in 15d 4;10h the increment in double elongation amounts to 369;57,29° = 2· (15d 4;10° · 12;11,27°/d). Subtracting one quantity from the other we obtain 37;29° (= 407;26,18° – 369;57,29°), in full agreement with the given entry.

The procedure is analogous for the other two quantities: in 15d 4;10h the Moon progresses in longitude 199;56,0° at the standard mean motion of 13;10,35°/d. As the longitude of the Moon for Ferrara at the Incarnation was 2,2;0,43°, 15d 4;10h earlier it was 4,42;5° (= 122;0,43° – 199;56,0°), in agreement with the entry in Table 12. Similarly, for the argument of lunar latitude, in 15d 4;10h the Moon progresses 200;44,18° at the standard mean motion of 13;13,46°/d. The argument of lunar latitude for Ferrara at the Incarnation was 3,33;42,24° = 213;42,24°;[10] hence, 15d 4;10h earlier it was 12;58° (= 213;42,24° – 200;44,18°). This result differs from the entry in Table 15 by 0;14°, but we note that the entry for the radix of the argument of lunar latitude at conjunction (5,52;52°: see Table 58) also differs by 0;14° from the recomputed value, indicating that the two radices for the argument of lunar latitude are consistent with each other.

As was the case for the radices of the Sun and the lunar node, the entries for the radices in 1400 and 1440 in Table 15 are obtained by adding those for the Incarnation to those corresponding to 1400

---

[8] See Ratdolt, *Tabule astronomice Alfontij*, d1v; see also Poulle, *Les tables alphonsines*, 128.

[9] See Ratdolt, *Tabule astronomice Alfontij*, d1v; see also Poulle, *Les tables alphonsines*, 127.

[10] See Ratdolt, *Tabule astronomice Alfontij*, d1v; see also Poulle, *Les tables alphonsines*, 128.

and 1440 years in the tables of the mean motions. Thus, for 1400 we obtain:

> 7d 15;51h = 15d 4;10h (Incarn.) + 20d 1;0h (1400y) – 27d 13;19h
> (1 anom. month),
> 178;56° = 37;29° (Incarn.) + 189;38° (1400y) + 311;49° (1 anom.
> month),
> 10;30° = 4,42;5° (Incarn.) + 1,25;21° (1400y) + 3;4° (1 anom. month),
> 3,0;12° = 13;12° (Incarn.) + 2,42;28° (1400y) + 4;32° (1 anom. month),

in agreement with the tabulated values.

Table 16. Monthly mean motion of the Moon

Na, f. 19r: *Menses non bisextiles* and *Menses bisextiles*
Nu, f. 23v: *Menses non bisextiles* and *Menses bisextiles*
Va, f. 190v: *Menses non bisextiles* and *Menses bisextiles*
Ed. 1495, f. b8r: *Tabula mensium non bisextilium* and *Tabula mensium bisextilium*
Ed. 1526, f. 16r: *Tabula mensium non bisextilium* and *Tabula mensium bisextilium*

This table has two sub-tables, one for a common year and another for a leap year, both beginning in January: see Table 16. The layout of this table is the same as that of Table 12, but here the first column lists the names of the 12 months. As was the case in previous tables, Regiomontanus omitted the column for the argument of lunar latitude.

Table 16: Monthly mean motion of the Moon for a
common year (excerpt)

Months	Time (d, h)		Double elongation (°)	Longitude (°)	Arg. latitude (°)
January	3	10;41	311;49	3; 4	4;32
February	3	21;23	263;39	6; 8	9; 2
...					
November	3	8;17	141;51	36;50	54;21
December	6	18;59	93;41	39;54	58;53

It is easy to check that the entry for the time for January results from subtracting the mean anomalistic month (27d 13;18,36h) from 31 days (January), but the entries for the three other quantities do not strictly correspond to a full month of January but only to a (shorter) anomalistic month, as can be seen in the first row of Table 14, where the entries are almost the same. The entries for December in a common year are: 6d 18;59h (time), 93;41° (double elongation), 39;54° (lunar longitude), and 58;53° (argument of lunar latitude). In the sub-table for a leap year, the entries for December are: 7d 18;59h (time), 93;41° (double elongation), 39;54° (lunar longitude), and 58;53° (argument of lunar latitude). It is readily seen that the entries for the last three quantities are almost the same as those for one year in Table 10, but we note that even though the entry for time has an extra day, the entries for the Moon remain the same for common and leap years.

Table 17. True positions of the Moon

Na, ff. 19v–22r: *Tabula lune*
Nu, ff. 24r–29v: *Incipiunt tabulae veri motus lunae*
Va, ff. 191r–193v: *Tabula lunae*
Ed. 1495, ff. b8v–d1r; no title
Ed. 1526, ff. 16v–25r: no title

In this double argument table, the vertical argument is the time, $t - t_0$, since the mean anomaly was 0° (where t is the time in question, and t_0 is the time when the mean anomaly was most recently 0°), expressed in integer days from 0d to 27d, whereas the horizontal argument is the double elongation from 0° to 350°, at intervals of 10°. As shown in Table 17, for each day, $t - t_0$, and each value of the double elongation, 2η, we are given four entries: the increment in true lunar longitude (in physical signs, degrees, and minutes), $\lambda'(t - t_0, 2\eta)$; lunar equation (in minutes and seconds), $q(t - t_0, 2\eta)$; hourly velocity of the Moon (in minutes and seconds), $v(t - t_0, 2\eta)$; and true argument of lunar latitude (in physical signs, degrees, and minutes), $\omega(t - t_0, 2\eta)$. In Regiomontanus's copy there is no column for the argument of lunar latitude.

Table 17: True position of the Moon (excerpt)

	0°			
Dies	Locus (°)	Equatio (°)	ad ho. (°/h)	Arg. lati. (°)
0	0; 0	0; 0,42	0;29,40	0; 0, 0
1	11;52	0; 0,54	0;29,30	0;11,55
2	23;40	0; 1, 6	0;29,30	0;23,46
3	35;28	0; 1,24	0;29,32	0;35,38
4	47;17	0; 1,18	0;30,15	0;47,30
5	59;23	0; 0,48	0;31, 7	0;59,39
6	1,11;50	0; 0,18	0;32,17	1;11, 9
7	1,24;43	0; 0, 6	0;32,57	1;25, 5
		adde		
8	1,37;56	0; 0,15	0;33,45	1;38,21
9	1,51;26	0; 1,18	0;35,10	1;51,55
10	2, 5;30	0; 1,24	0;36,18	2; 6, 2
11	2,20; 1	0; 1,24	0;36,57	2;20,36
12	2,34;48	0; 1,24	0;37,12	2;35,26
13	2,49;41	0; 1, 6	0;37, 0	2;50,22
14	3, 4;29	0; 0,48	0;36,48	3; 5,13
15	3,19;12	0; 0,54	0;36,32	3;20, 0
16	3,33;49	0; 1, 6	0;36,22	3;34,40
17	3,48;22	0; 1,18	0;35,57	3;49,16
18	4, 2;45	0; 1,30	0;34,57	4; 3,42
19	4,16;44	0; 1,48	0;34, 2	4;17,44
20	4,30;21	0; 1,24	0;33,10	4;31,25
21	4,43;37	0; 1,12	0;32,42	4;44,44
22	4,56;42	0; 1, 6	0;31,58	4;57,52
23	5, 9;29	0; 0,30	0;31, 2	5;10,42
		minue		
24	5,21;54	0; 0,36	0;29,52	5;23,10
25	5,33;51	0; 1, 6	0;29,10	5;35,10
26	5,45;31	0; 1, 6	0;29, 7	5;46,54
27	5,57;10	0; 1, 0	0;29,17	5;58,36

Table 17 (*cont.*)

	10°			
Dies	Locus (°)	Equatio (°)	ad ho. (°/h)	Arg. lati. (°)
0	5,59;53	0; 0,42	0;29,35	5;59,53
1	11;43	0; 0,48	0;29,25	0;11,46
2	23;29	0; 1, 6	0;29,22	0;23,35
3	35;14	0; 1,18	0;29,35	0;35,24
4	47; 4	0; 1, 6	0;30,28	0;47,17
5	59;15	0; 0,42	0;31,20	0;59,31
6	1,11;47	0; 0,20	0;32,18	1;12, 6
		adde		
7	1,24;42	0; 0,42	0;33,13	1;25, 4
8	1,37;59	0; 0,48	0;34,10	1;38,24
9	1,51;39	0; 0,54	0;35,13	1;52, 8
10	2, 5;44	0; 1,24	0;36,18	2; 6,16
11	2,20;15	0; 1,24	0;37, 0	2;20,50
12	2,35; 3	0; 1, 6	0;37, 2	2;35,41
13	2,49;52	0; 0,54	0;36,53	2;50,33
14	3, 4;37	0; 0,54	0;36,50	3; 5,21
15	3,19;21	0; 0,54	0;36,37	3;20, 9
16	3,34; 0	0; 1,24	0;36,27	3;34,51
17	3,48;35	0; 1,42	0;36, 2	3;49,29
18	4, 3; 0	0; 1,30	0;35, 5	4; 3,57
19	4,17; 2	0; 1, 6	0;33,23	4;18, 2
20	4,30;35	0; 1,18	0;33, 5	4;31,39
21	4,43;49	0; 0,54	0;32,40	4;44,56
22	4,56;53	0; 0,18	0;31,42	4;58, 6
		minue		
23	5, 9;34	0; 0, 6	0;30,35	5;10,47
24	5,21;48	0; 0,48	0;29,40	5;23, 4
25	5,33;40	0; 1,18	0;29,10	5;34,59
26	5,45;20	0; 1,12	0;29,10	5;46,43
27	5,57; 0	0; 0,48	0;29,29	5;58,26

In an anomalistic month of 27d 13;18,36h the Moon, initially at apogee, returns to it and travels 6 physical signs; the intermediary tabulated values of $\lambda'(t - t_0, 2\eta)$ represent the increment of the true lunar longitude at the beginning of each of the 27 days listed. This increment is not counted from Aries 0°; rather, it is an increment over the true longitude at the time when the mean anomaly was 0°. The true argument of latitude, $\omega(t - t_0, 2\eta)$, is related to the increment in true longitude by means of the expression:

$$\omega(t - t_0, 2\eta) = \lambda'(t - t_0, 2\eta) - \lambda_N,$$

where λ_N is the longitude of the lunar node which, in Alfonsine astronomy, moves at the rate of $-0;3,10,38°/d$. Indeed, this is the quantity that can be deduced from the table where, for any ω and λ', the following relation holds:

$$\omega(t - t_0, 2\eta) = \lambda'(t - t_0, 2\eta) - (t - t_0) \cdot 0;3,10,38.$$

By inspection of the table we see that the tabulated lunar equation was derived from the lunar longitude according to the expression:

$$q(t - t_0, 2\eta) = [\lambda'(t - t_0, 2\eta) - \lambda'(t - t_0, 2\eta + 10°)]/10,$$

and this is why all entries for the lunar equation in the double argument table are multiples of 6. Thus, for example,

$$q(1,0) = [\lambda'(1, 0) - \lambda'(1, 10)]/10 = (11;52° - 11;43°)/10 = 0;9°/10$$
$$= 0;0,54°.$$

Similarly, the tabulated hourly velocity results from the following expression:

$$v(t - t_0, 2\eta) = [\lambda'(t - t_0 + 1d, 2\eta) - \lambda'(t - t_0, 2\eta)]/24.$$

Thus, for example,

$$v(1, 0) = [\lambda'(2, 0) - \lambda'(1, 0)]/24 = (23;40° - 11;52°)/24 = 11;48°/24$$
$$= 0;29,30°.$$

Both these quantities are used for interpolation purposes: $q(t - t_0, 2\eta)$ is for interpolation between consecutive days in an anomalistic month to take care of the hours and minutes at the time under consideration, and $v(t - t_0, 2\eta)$ is for interpolation between consecutive sets of entries for double elongation (at intervals of $10°$) at the time under consideration. The combined effect of these two factors is to be added both to the increment in lunar longitude and to the argument for lunar latitude derived from the table.

Chapter 11 provides a worked example for the use of the lunar tables for 1447, July 18, at 12;30h. Although not stated explicitly, the general method to compute the true position of the Moon (true longitude and true argument of latitude) by means of Bianchini's tables consists, first, in determining in the tables for mean motions the mean lunar anomaly, the double elongation, the mean lunar longitude, and the mean argument of lunar latitude. The key move for Bianchini is to determine the double elongation at the time when the mean anomaly was most recently equal to 0°. To facilitate this computation Bianchini set up these special tables for determining the time when the mean lunar anomaly is 0°. Then one enters into the double argument table with time (mean anomaly) and the double elongation at the time when the mean anomaly was most recently equal to 0° and looks for the true lunar longitude and the true argument of lunar latitude closest to the given mean anomaly and mean double elongation. Next, one interpolates in the double argument table to take into account the excesses of time (anomaly) and double elongation. The interpolation is greatly facilitated by the tabulated quantities. In the worked example, we are first told to find in Tables 12, 13, and 16 the entries for 1440y, 6y, and June for the four quantities: time, double elongation, lunar longitude, and argument of lunar longitude. We are also told to add 18d 12;30h to time. We note again that Bianchini uses complete days. After adding the corresponding radices for the Incarnation found in Table 15, the resulting values are:

time = 89d 16;23h = 26d 3;1h (1440y) + 14d 4;33h (6y) + 15d 16;9h (June) + 18d 12;30h (date) + 15d 4;10h (Incarn.)
double elongation = 117;16° = 214;58° (1440y) + 153;53° (6y) + 70;56° (June) + 37;29h (Incarn.)
mean longitude = 1,35;20° = 4,32;19° (1440y) + 4,2;31° (6y) + 18;25° (June) + 4,42;5h (Incarn.)
mean argument of latitude = 1,20;55° = 42;46° (1440y) + 5,57;47° (6y) + 27;10° (June) + 13;12h (Incarn.).

We are then told to subtract from the time shown above the entry corresponding to 3 anomalistic months in Table 14, and to add to the other quantities shown above the corresponding entries for 3 anomalistic months. Note that, as a result of the organization Bianchini gave to his tables, one has to subtract from the time the entry for 3 anomalistic months. The reason is that in 3 anomalistic months the anomaly advances by 3 · 13;3,53,57° · (27d 13;18,36h) = 1080;0° = 0°; hence, the motion in anomaly in 3 anomalistic months can be disregarded. But in the time corresponding to 3 anomalistic months the increments in the other quantities are not 0° and so their values have to be added. The resulting values are:

time = 7d 0;27h = 89d 16;23h – 82d 15;56h (3 anom. months)
double elongation = 332;44° = 117;16° + 215;28° (3 anom. months)
mean longitude = 1,44;33° = 1,35;20° + 9;13° (3 anom. months)
mean argument of latitude = 1,34;30° = 1,20;55° + 13;35° (3 anom. months).

As we said above, let t be the time in question and t_0 the time when the mean anomaly was most recently 0°; in order to use Table 17 we need to find $t - t_0$, at which time the mean anomaly was the same as it was at t. We then find the double elongation at $t - t_0$, for it is needed as the second argument in the table. In this case, $t - t_0$ = 7d 0;27h, and the double elongation, 2η = 332;44°. Moreover, the mean longitude at $t - t_0$ is $\bar{\lambda}_0$ = 1,44;33°, and the mean argument of latitude is $\bar{\omega}$ = 1,34;30°. Since the arguments in Table 17 are given in integer days and increments of 10° of double elongations, we will have to interpolate.

Thus, one enters Table 17 with $t - t_0$ = 7d and 2η = 330°. The corresponding entries in the double argument table are:

λ'(7, 330) = 1,25;0°
q(7, 330) = 0;0,48°
v(7, 330) = 0;32,40°/h
ω(7, 330) = 1,25;22°.

Next we are told to take the excess of time over 7d (0;27h) and the excess of double elongation over 330° (2;44°) and multiply them by v(7, 330) and q(7, 330), respectively. The result is

0;12,19° = 0;27h · 0;32,40°/h – 2;44° · 0;0,48° (0;12,18°, if correctly computed).

This amount has to be added both to λ'(7, 330) and ω(7, 330), to obtain λ'($t - t_0$, 2η) = λ'(7d 0;27h, 332;44°) = 1,25;12,19° (= 1,25;0° +0;12,19°) and ω'($t - t_0$, 2η) = ω'(7d 0;27h, 332;44°) = 1,25;34,19° (= 1,25;22° + 0;12,19°), respectively. Finally, the true lunar longitude (on the 9th sphere) at the time sought is:

3,9;45,19° = 1,25;12,19° + 1,44;33°,

that is,

$\lambda = \lambda'(t - t_0, 2\eta) + \bar{\lambda}$,

where $\bar{\lambda}$ = 1,44;33°, the mean longitude found before, plays the role of an intermediary radix. Similarly, the true argument of lunar latitude, ω, is:

$3,0;4° = 1,25;34,19° + 1,34;30°,$

that is,

$\omega = \omega'(t - t_0, 2\eta) + \bar{\omega},$

where $\bar{\omega} = 1,34;30°$, the mean argument of lunar latitude found before, plays the role of an intermediary radix.

Using the standard Alfonsine Tables, we have recomputed the entry for $t - t_0 = 7d$ and $2\eta = 330°$ in this table, $\lambda'(7, 330) = 1,25;0°$ (= 85;0°). After 190 days, the increment in double elongation is $170;40,18°$ (= 7d · 2 · 12;11,27°/d) and the increment in anomaly is $91;27,18°$ (= 7d · 13;3,54°/d). Thus, the double elongation, 2η, is $170;40° + 330° = 140;40°$. The equation of center, $c_3(2\eta)$, corresponding to $140;40°$, is $+11;5°$; and the value for the minutes of proportion, $c_4(2\eta)$, is 52. To recompute this entry where $t - t_0$ (7d in our case) is the time elapsed since the instant when the mean anomaly was $0°$, we need to use the true anomaly, $\alpha = 102;32°$ (= 91;27° + 11;5°). The equation of argument, $c_6(\alpha)$, corresponding to $102;32°$, is $-4;53°$, and the correction to this equation, $c_5(\alpha)$, is listed as $-2;39°$. According to the rules for computing a lunar position with the standard Alfonsine Tables, we have to compute:

$\lambda'(t - t_0, 2\eta) = \Delta\bar{\lambda} + c_6(\alpha) + c_4(2\eta) \cdot c_5(\alpha),$

where c_i refers to a value computed with interpolation from the entries in column i in the standard Alfonsine Tables. In Fig. 2, the observer is at O, the center of the epicycle at C, the mean epicyclic apogee is at $\bar{A}_e$ (directed towards D'), the true epicyclic apogee is at A_e, and the Moon is at L. There are two equations: q_1 is approximated by $c_6(\alpha)$, and q_2 is approximated by $c_4(2\eta) \cdot c_5(\alpha)$.

Now, after 7 days, the increment in longitude, $\Delta\bar{\lambda}$, is $7d \cdot 13;10,35° = 92;14,5°$. Thus, the total increment in longitude, λ', is $85;3°$, for:

$\lambda'(7, 330) = 92;14° - 4;53° - 0;52 \cdot 2;39° = 92;14° - 4;53° - 2;18° = 85;3°,$

in close agreement with the given entry (85;0°). All our other recomputations with the standard Alfonsine Tables also yield very good agreement, often exactly to the minute.

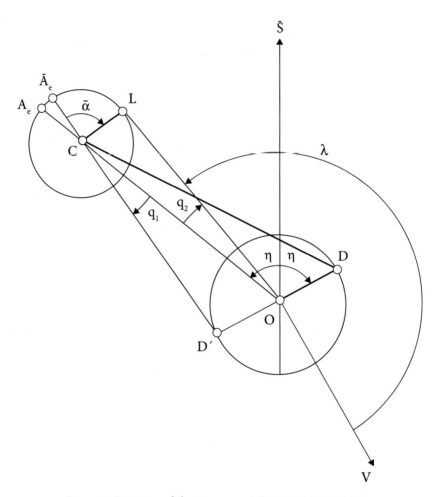

Figure 2: Position of the Moon on July 18, 1447 at 12;30h.

Table 18. Daily mean motion of the Moon

Na, f. 23v: *Tabula motus lune*
Va, f. 195v: *Tabula motus lunae*
Ed. 1495, f. d1v: *Tabula motus lune in diebus*
Ed. 1526, f. 25v: *Tabula motus* [symbol for the Moon] *in diebus*

In this table there are four columns: see Table 18. The first gives the argument, the number of days from 1 to 30 at intervals of 1 day. The entries in the other columns represent the longitude of the Moon, the lunar anomaly, and the argument of lunar latitude, and the entries for 1 day are all approximations of the parameters used in Alfonsine astronomy.

Table 18: Daily mean motion of the Moon (excerpt)

Days	Longitude (°)	Anomaly (°)	Arg. latitude (°)
1	13;10,35	13; 3,54	13;13,46
2	26;21,10	26; 7,48	26;17,31
...			
29	22; 6,56	18;53, 5	23;39, 4
30	35;17,31	31;56,59	36;52,50

Table 19. Hourly mean motion of the Moon

Na, f. 23v: *Tabula motus lune*
Va, f. 196r: *Ad horas lunae*
Ed. 1495, f. d2r: *Tabula motus lune in horis*
Ed. 1526, f. 26r: *Tabula motus* [symbol for the Moon] *in horis*

This table has the same layout as Table 18, but the argument here is the number of hours, from 1 to 30 at intervals of 1 hour: see Table 19. The entries for 24 hours are the same as those for 1 day in Table 18.

Table 19: Hourly mean motion of the Moon (excerpt)

Hours	Longitude (°)	Anomaly (°)	Arg. latitude (°)
1	0;32,56	0;32,40	0;33, 4
2	1; 5,53	1; 5,19	1; 6, 9
...			
29	15;55,17	15;47,13	15;59, 7
30	16;38,13	16;19,52	16;32,12

Table 20. Lunar latitude

Na, f. 22v: *Tabula latitudinis lune*
Nu, f. 30r: *Tabula latitudinis lunae*
Va, f. 53r: *Tabula latitudinis lunae*
Ed. 1495, f. d2v: *Tabula latitudinis lune*
Ed. 1526, f. 26v: *Tabula latitudinis lune*

The latitude of the Moon, β, is displayed in degrees, minutes, and seconds, as a function of the argument of latitude, ω, at intervals of 1 degree: the maximum entry is 5;0,0° for arguments of latitude 90° and 270°. This is the same table as in ed. 1483 of the Alfonsine Tables,[11] but for a few entries, indicated in Table 20A.

Table 20A: Lunar latitude

Arg. latitude (°)	Bianchini (°)	Alf. T. 1483 (°)
...		
32	2;38,52	2;38,52
33	2;43,17	2;43,<u>57</u>
34	2;47,39	2;47,39
...		
56	4; 8,37	4; 8,37
57	4;11,30	4;11,<u>34</u>
58	4;14,22	4;14,22
...		
63	4;27,14	4;27,14
64	4;29,34*	4;29,34
65	4;31,49	4;31,49
...		

* Both editions of 1495 and 1526 have 4;29,24°.

---

[11] Ratdolt, *Tabule astronomice Alfontij*, h1r.

The same table is already found in the zij of al-Battānī,[12] and Bianchini followed the tradition of his predecessors.

As was the case for the equation of access and recess of the 8th sphere (Table 1), the entries in this table were computed using a formula in spherical, not plane, trigonometry, which can be written in modern terms as

$$\sin \beta = \sin 5 \cdot \sin \omega,$$

or its equivalent

$$\beta = \arcsin (\sin 5 \cdot \sin \omega).$$

In particular, for $\omega = 30°$, we find $\beta = 2;29,51°$ (text: $2;29,52°$), and for $\omega = 60°$, we obtain $\beta = 4;19,44°$ (text: $4;19,47°$). The recomputed entries for $\omega = 33°, 57°$, and $64°$ are $2;43,15°$, $4;11,30°$, and $4;29,34°$, indicating that both printed editions of the tables of Bianchini have an isolated error in the entry for $\omega = 64°$, whereas the *editio princeps* of the Alfonsine Tables has isolated errors for $\omega = 33°$ and $\omega = 57°$.

Other astronomers used the same formula to compute lunar latitude with a different parameter for the maximum value: see, e.g., Levi ben Gerson for whom the maximum lunar latitude is $4;30°$.[13]

Mean motion of the planets in collected years: Tables 21 (Saturn), 28 (Jupiter), 35 (Mars), 42 (Venus), and 49 (Mercury)

Na, f. 23r–v: *Tabula radicum planetarum* and *Tabula radicum Mercurii*
Nu, ff. 30v, 31v: *Tabula radicum planetarum* and *Mercurii*
Va, ff. 194r–195r: *Tabula radicum planetarum* and *Tabula radicum Mercurii*
Ed. 1495, f. d3r (Saturn: *Tabula radicum Saturni in annis collectis*); f. g7r (Jupiter); f. l3r (Mars); f. s2r (Venus); and f. çlr (Mercury)
Ed. 1526, f. 27r (Saturn: *Tabula radicum* [symbol for Saturn] *in annis collectis*); f. 55r (Jupiter); f. 83r (Mars); f. 138r (Venus); and f. 193r (Mercury)

The quires, i.e., gatherings of 8 folios, in ed. 1495 are labeled a to z, followed by three unusual symbols that we represent by &, ç, and r̂, and then by capital letters from A to O: a list of labels for the quires appears on f. D6r at the end of this edition. The three unusual symbols are the

[12] Nallino, *Al-Battānī*, 2:78–83.
[13] Goldstein, *Astronomical Tables of Levi ben Gerson*, 132.

standard abbreviations in Latin for the conjunction 'et', the prefix 'con', and the ending 'rum', all of which occur in this edition. Presumably, the printer chose these symbols because he had them in his lower case font.

In each of these tables there are three columns: see Table 21. The first is for the argument, collected years from 40 to 2000 at intervals of 40y. The entries in the other columns represent the time (in days, hours, and minutes), here called *superatio*, and the increment in longitude (superior planets) or in anomaly (inferior planets), in physical signs, degrees, and minutes. Both these increments are here called *motus* and correspond to the number of collected years. The time displayed in column 2 is the excess over an integer number of periods of anomaly, which are explicitly given in Chapter 14 (see Table 24A). We note that Regiomontanus replaced what Bianchini designated as *superatio* by the term *argumenti*.

Table 21: Mean motion of Saturn in collected years (excerpt)

Years	Superatio (d, h)		Motus (°)
40	242	12;16	2, 1;16
80	106	22;19	4,15;12
...			
1400	169	20;43	3,22;53
...			
2000	26	14;37	5,48;30

MSS Na and Nu have a single table for Saturn, Jupiter, Mars, and Venus, and a separate one for Mercury.

Mean motion of the planets in expanded years: Tables 22 (Saturn), 29 (Jupiter), 36 (Mars), 43 (Venus), and 50 (Mercury)

Na, f. 23r–v: *Tabula radicum planetarum* and *Tabula radicum Mercurii*
Nu, f. 31r–v: *Tabula radicum planetarum* and *Mercurii*
Ed. 1495, f. d3r–v (Saturn: *Tabula radicum* [symbol for Saturn] *in annis collectis*); f. g7r–v (Jupiter); f. l3r–v (Mars); f. s2r–v (Venus); and f. çlr–v (Mercury)

Ed. 1526, f. 27r–v (Saturn: *Tabula radicum* [symbol for Saturn] *in annis collectis*); f. 55r–v (Jupiter); f. 83r–v (Mars); f. 138r–v (Venus); and f. 193r–v (Mercury)

The layout of this table is the same as that of Tables 21, 28, 35, 42, and 49, but here the argument is the number of expanded years, from 1y to 40y: see Table 22. The entries for 1y are 365d 0;0h (*superatio*) and 0;0° (*motus*) for Saturn, Jupiter, Mars, and Venus; and 17d 8;45h (*superatio*) and 5,42;39° (*motus*) for Mercury. This surprising feature comes from the special meaning given by Bianchini to the term *motus*: the increment in longitude or anomaly after one or more periods of anomaly. That is, when a planet completes its period (e. g., 378d 2;12h in the case of Saturn: see Table 24A) the time is decreased by this quantity, so that the *superatio* is always less than it. In contrast, the *motus* associated with one period of anomaly (12;40° for Saturn) has to be added. Thus, in the present table, where for 1 year we are given 365d 0;0h and 0;0° for four planets, there is nothing to add to the longitude, for their periods of anomaly exceed 365d 0;0h. Now, in the case of Saturn, the entries for 2 years are 351d 21;48h (*superatio*) and 12;40° (*motus*). For Bianchini, this means that at the end of the second year, when 730d 0;0h (= 2 · 365d 0;0h) have elapsed, the excess in time is 351d 21;48h (= 730d 0;0h – 378d 2;12h) and the entry for the *motus* is that of one period of anomaly (12;40°: see Table 24A, and not the motion corresponding to 351d 21;48h) because only one period of anomaly has been completed.

Table 22: Mean motion of Saturn in expanded years (excerpt)

Years	Superatio (d, h)		Motus (°)
1	365	0; 0	0, 0; 0
2	351	21;48	0,12;40
3	338	19;36	0,25;20
...			
40	242	12;16	2, 1;16

The entries for 40 expanded years for each planet agree with those for 40 collected years in Tables 21, etc., respectively. Na, Nu, and Va have a single table for Saturn, Jupiter, Mars, and Venus, and a separate one for Mercury.

Radices of the planets: Tables 23 (Saturn), 30 (Jupiter), 37 (Mars), 44 (Venus), and 51 (Mercury)

Na, f. 23r–v: *Radices*
Nu, f. 31r–v: *Tabula radicum planetarum* and *Mercurii*
Va, ff. 194v–195r: *Tabula radicis Christi* and *Radices Mercurii*
Ed. 1495, f. d3v (Saturn: *Radices* [symbol for Saturn]; f. g7v (Jupiter);
    f. l3v (Mars); f. s2v (Venus); and f. çlv (Mercury)
Ed. 1526, f. 27v (Saturn: *Radices* [symbol for Saturn]); f. 55r (Jupiter);
    f. 83v (Mars); f. 138v (Venus); and f. 193v (Mercury)

Again, the layout is the same as that of Tables 21, etc. and 22, etc. We are given the radices for the Incarnation, 1400, and 1440. Ed. 1526 adds entries for 1480 and 1500 (all planets), as well as for 1508 (Saturn), 1496 (Mars), and 1510 (Mercury). In Table 23A we only display the entries for the Incarnation, 1400, and 1440.

Table 23A: Radices of the planets

	Incarnation			1400			1440		
	Superatio (d, h)		*Motus* (°)	*Superatio* (d, h)		*Motus* (°)	*Superatio* (d, h)		*Motus* (°)
Saturn	214	11;13	1, 6;54	6	5;44	4,42;26	248	18; 0	43;42
Jupiter	108	5;24	2,51;37	89	15;53	3,20;12	339	20;37	5,13;54
Mars	513	5;40	2,12;25	227	14;38	1, 4;14	18	21;29	4,30;15
Venus	209	18;40	1,11;32	39	23;47	4, 9;10	51	19;46	3,57;48
Mercury	14	13;19	4,23;57	108	15; 3	3, 1;31	2	1;22	4,46;51

Na, Nu, and Va have a single table for Saturn, Jupiter, Mars, and Venus, and a separate one for Mercury.

The entries for *superatio* for the Incarnation are to be understood as meaning that the anomaly of each planet was 0° at this time before the Incarnation, respectively. For example, the anomaly of Saturn was 0° at 214d 11;13h before the Incarnation. In order to justify this claim, let us consider two quantities, both for the Incarnation and the longitude of Ferrara: the mean solar longitude (4,38;17,44°: see Table 5) and the mean longitude of Saturn (1,14;5,13°).[14] Their difference is the mean

---

[14] See Ratdolt, *Tabule astronomice Alfontij*, d2r; see also Poulle, *Les tables alphonsines*, 128.

anomaly at the Incarnation (3,24;12,31°). Now, the daily mean motion in anomaly is the difference between the daily solar motion and the daily motion of the planet: 0;59,8,19,37 − 0;2,0,35,18 = 0;57,7,44,19°/d. Therefore, we find 3,24;12,31/0;57,7,44,19 ≈ 214;28,16d = 214d 11;18h, in agreement with the entry in the table. During this time interval, the planet progresses in longitude 214d 11;13h · 0;2,0,35,18°/d ≈ 7;11°. Finally, the longitude of the planet at the beginning of the period of anomaly that includes the Incarnation was 1,14;5° − 7;11° = 1,6;54°, also in agreement with the entry in the text. The rest of the entries for the Incarnation are found analogously.

As was the case for the radices of the Sun and the lunar node (Table 5) and the Moon (Table 15), the entries for the radices in 1400, 1440,…, in Tables 23, etc. are obtained by adding those for the Incarnation to those corresponding to 1400, 1440,… years in the tables of the mean motions. Thus, for 1400 we find for Saturn:

6d 5;44h = 214d 11;13h (Incarn.) + 169d 20;43h (1400y) − 378d 2;12h (1 anom. period),

and

4,42;27° = 1,6;54° (Incarn.) + 3,22;53° (1400y) + 12;40° (1 anom. period),

in agreement with the tabulated values, and similarly for the other planets.

Multiples of periods of anomaly: Tables 24 (Saturn), 31 (Jupiter), 38 (Mars), 45 (Venus), and 52 (Mercury)

Na, f. 23r–v: *Tabula brevis*
Nu, f. 31r–v: *Tabula brevis*
Va, ff. 194v–195r: *Tabula brevis*
Ed. 1495, f. d3v (Saturn: *Tabella brevis*); f. g7v (Jupiter); f. l3v (Mars); f. s2v (Venus); and f. çlv (Mercury)
Ed. 1526, f. 27v (Saturn: *Tabula brevis*); f. 55r (Jupiter); f. 83v (Mars); f. 138v (Venus); and f. 193v (Mercury)

Each of these tables has 4 rows for successive multiples of the appropriate periods of anomaly. In each case there are entries for the corresponding increments in longitude (*motus*). In Table 24A we display the first row of each of these tables. The entries for the periods of anomaly of each planet appear as a small table in Chapter 14.

Table 24A: Periods of planetary anomaly

	Period (d, h)		*Motus* (°)
Saturn	378	2;12	12;40
Jupiter	398	21;12	33; 9
Mars	779	22;23	48;44
Venus	583	22;14	3,35;33
Mercury	115	21; 5	1,54;13

Na, Nu, and Va have a single table for Saturn, Jupiter, Mars, and Venus, and a separate one for Mercury.

These periods are not explicitly stated in most of the manuscripts and printed editions of the Alfonsine tables we have consulted, but they can easily be derived from the data usually given in them:

378d 2;12h = 378;5,30d ≈ 360°/(0;59,8,19,37 °/d − 0;2,0,35,18°/d),
398d 21;12h = 398;53,0d ≈ 360°/(0;59,8,19,37°/d − 0;4,59,15,27°/d),
779d 22;23h = 779;55,57,30d ≈ 360°/(0;59,8,19,37°/d − 0;31,26,38,40°/d),
583d 22;14h = 583;55,35d ≈ 360°/0;36,59,27,23°/d,
115d 21;5h = 115;52,42,30d ≈ 360°/3;6,24,7,43°/d,

where 0;59,8,19,37°/d is the mean motion in longitude of the Sun, Venus, and Mercury; 0;2,0,35,18°/d, 0;4,59,15,27°/d, and 0;31,26,38,40°/d are the mean motions in longitude of Saturn, Jupiter, and Mars, respectively; and 0;36,59,27,23°/d and 3;6,24,7,43°/d are the mean motions in anomaly of Venus and Mercury, respectively. All these parameters are used in the standard Parisian Alfonsine Tables.

The entries for the *motus* are also derived from the same parameters of Alfonsine astronomy:

12;40° ≈ 378d 2;12h · 0;2,0,35,18°/d,
33;9° ≈ 398d 21;12h · 0;4,59,15,27°/d,
48;44° ≈ (779d 22;23h · 0;31,26,38,40°/d) − 360°,
3,35;33° ≈ (583d 22;14h · 0;59,8,19,37°/d) − 360°,
1,54;13° ≈ 115d 21;5h · 0;59,8,19,37°/d.

Table 53. Monthly mean motion of Mercury in anomaly

Na, f. 23v: no title
Nu, f. 31v: *Menses Mercurii*
Va, f. 195r: no title

Ed. 1495, f. çlv: *Tabula medii motus* [symbol for Mercury] *in mensibus*
Ed. 1526, f. 193v: *Tabula medii motus et argumenti* [symbol for Mercury] *in mensibus*

There are no such tables for the rest of the planets, because Mercury is the only planet for which the period of anomaly does not exceed 1 solar year. This table is used both for a common year and a leap year, beginning in January: see Table 53. The first column lists the names of the 12 months. The second and third columns display the excess over an integer number of synodic periods of Mercury (115d 21;5h) for each month of a common and a leap year, respectively, and the fourth column gives the increments in anomaly corresponding to these times. The period used here is synodic, because this is the period of anomaly for Mercury.

Table 53: Monthly motion of Mercury in anomaly

	Common year (d, h)		Leap year (d, h)		*Motus* (°)
January	31	0; 0	31	0; 0	0; 0
February	59	0; 0	60	0; 0	0; 0
March	90	0; 0	91	0; 0	0; 0
April	4	2;55	5	2;55	1,54;13
May	35	2;55	36	2;55	1,54;13
June	65	2;55	66	2;55	1,54;13
July	96	2;55	97	2;55	1,54;13
August	11	5;50	12	5;50	3,48;26
September	41	5;50	42	5;50	3,48;26
October	72	5;50	73	5;50	3,48;26
November	102	5;50	103	5;50	3,48;26
December	17	8;45	18	8;45	5,42;39

Daily mean motion of the planets: Tables 25 (Saturn), 32 (Jupiter), 39 (Mars), 46 (Venus), and 54 (Mercury)

Na, f. 23v: *Tabula medius motus planetarum et argumenti*
Va, f. 195v: *Tabula medius motus planetarum et argumenti*
Ed. 1495, f. d3v (Saturn: *Tabula medii motus* [symbol for Saturn] *in diebus*); f. g7v (Jupiter); f. l3v (Mars); f. s2v (Venus); and f. çlv (Mercury)
Ed. 1526, f. 27v (Saturn: *Tabula medii motus* [symbol for Saturn] *in diebus*); f. 55r (Jupiter); f. 83v (Mars); f. 138v (Venus); and f. 193v (Mercury)

This table for each planet has two columns: see Table 25. The first column represents the argument: the number of days, in decreasing order, from 700d to 100d at intervals of 100d, from 100d to 10d at intervals of 10d, and from 10d to 1d at intervals of 1 day. The second column is the mean motion in longitude (in physical signs, degrees, minutes, and seconds). In this case Bianchini applies the standard meaning of *motus*, that is, an increment in longitude independent of the period of anomaly. Thus, in the case of Saturn, the entries for 700d to 400d are zero, and the first non-zero entry is for 300d (10;2,57°) and indeed this is the increment in longitude traveled by Saturn in 300d, for 10;2,57°/300d = 0;2,0,35°, which is an approximation of the parameter used in Alfonsine astronomy. The other planets are treated analogously, except for Mercury, which has a smaller period of anomaly. Therefore, for each planet, Tables 25, ctc., and those following them, namely, Tables 26, etc., are to be used only when the time does not exceed one period of anomaly. For times greater than these periods, Tables 21, etc., and 22, etc., with a different meaning of *motus*, apply. For Venus and Mercury the entries represent increments in anomaly.

Table 25: Mean motion of Saturn in days (excerpt)

Days	Mean Motion
(°)	
700	0; 0, 0
…	
400	0; 0, 0
300	10; 2,57
200	6;41,58
…	
2	0; 4, 1
1	0; 2, 1

Na and Va have a single table for all 5 planets and display an extra column for the Sun.

Hourly mean motion of the planets: Tables 26 (Saturn), 33 (Jupiter), 40 (Mars), 47 (Venus), and 55 (Mercury)

Na, f. 23v: *Tabula medius motus planetarum et argumenti*
Va, f. 196r: *Tabula ad horas et fractiones horarum*

Ed. 1495, f. d3v (Saturn: *Tabula medii motus* [symbol for Saturn] *in horis*); f. g7v (Jupiter); f. l3v (Mars); f. s2v (Venus); and f. çlv (Mercury)

Ed. 1526, f. 27v (Saturn: *Tabula medii motus* [symbol for Saturn] *in horis*); f. 55r (Jupiter); f. 83v (Mars); f. 138v (Venus); and f. 193v (Mercury)

This table for each planet has the same two columns as Tables 25, etc.: see Table 26. The first column lists the argument, the number of hours in increasing order from 1h to 30h at intervals of 1 hour. The values tabulated in the second column have the same meaning as in Tables 25, etc. The entries for 24 hours are the same as those for 1 day in Tables 25, etc. Na and Va have a single table for all 5 planets and display an extra column for the Sun.

Table 26: Mean motion of Saturn in hours (excerpt)

Hours	Mean Motion (°)
1	0; 0, 5
2	0; 0,10
...	
29	0; 2,26
30	0; 2,31

True positions of the planets: Tables 27 (Saturn), 34 (Jupiter), 41 (Mars), 48 (Venus), and 56 (Mercury)

Na, ff. 24r–32v (Saturn: *Saturnus*); ff. 33r–41v (Jupiter); ff. 42r–59v (Mars); ff. 60r–77v (Venus); and ff. 78r–86v (Mercury)

Nu, ff. 32r–40v (Saturn: *Incipiunt tabulae veri loci Saturni*), ff. 41r–49v (Jupiter); ff. 50r–67v (Mars); ff. 68r–85v (Venus); and ff. 86r–94v (Mercury)

Rc, f. 20r–v (Saturn: *Saturnus*; only one folio is preserved, the rest was cut off); ff. 21r–25r (Jupiter); ff. 25v–34r (Mars); ff. 34v–42r (Venus); and ff. 42v–46r (Mercury)

Va, ff. 196v–204r (Saturn: *Saturnus*); ff. 204v–213r (Jupiter); ff. 213v–231r (Mars); ff. 231v–249r (Venus); and ff. 249v–258r (Mercury)

Ed. 1495, ff. d4r–g6v (Saturn: *Residuum* [symbol for Saturn]); ff. g8r–l2v (Jupiter); ff. l4r–s1v (Mars); ff. s3r–&8v (Venus); and ff. çlr–C5v (Mercury).

Ed. 1526, ff. 28r–54v (Saturn: *Residuum* [symbol for Saturn]); ff. 56r–82v
(Jupiter); ff. 84r–137v (Mars); ff. 139r–192v (Venus); and ff. 194r–229v
(Mercury)

For each planet we are given a double argument table. The vertical argument is the time, in days, within an anomalistic period, at intervals of
10 days, or 3 days in the case of Mercury, from 0d to 378d (Saturn),
398d (Jupiter), 779d (Mars), 583d (Venus), and 115d (Mercury). For
special planetary positions the intervals are occasionally shorter. Thus,
the vertical argument represents mean anomaly. The horizontal argument is the 'center', from 0,0° to 5,50°, at intervals of 10°.

As shown in Tables 41 and 56, for each day and each value of the
'center', we are given eight entries, where $t - t_0$ is the number of days
since the mean anomaly was 0° and $\bar{\kappa}$ is the argument of center when the
mean anomaly was 0°; increment in longitude (in physical signs, degrees,
and minutes), $\lambda'(t - t_0, \bar{\kappa})$, here called *locus*; equation in longitude (in
minutes and seconds), $q_L(t - t_0, \bar{\kappa})$, here called *equatio*; daily velocity
in longitude (in minutes and seconds), $v_D(t - t_0, \bar{\kappa})$; hourly velocity in
longitude (in minutes and seconds), $v_H(t - t_0, \bar{\kappa})$; latitude (in degrees
and minutes), $\beta(t - t_0, \bar{\kappa})$; equation in latitude (in minutes and seconds),
$q_B(t - t_0, \bar{\kappa})$, also called *equatio*; elongation (in physical signs, degrees,
and minutes), $\eta(t - t_0, \bar{\kappa})$; and equation in elongation (in minutes and
seconds), $q_E(t - t_0, \bar{\kappa})$, also called *equatio*. For the *locus* it is indicated
whether it is *directus* or *retrogradus* and for the latitude whether
meri[dionalis] or *sept[entrionalis]*, i.e., south or north. The elongation
listed here is the difference between the true longitudes of the Sun and
the planet. For the superior planets the symbol for the Sun appears
under the heading elongation, but for the inferior planets this symbol
alternates with that of the corresponding planet. As explained in Chapter
20, the symbol for the Sun is associated with an inferior planet when it
is seen above the eastern horizon in the morning, whereas the symbol
for the planet is associated with an inferior planet when it is seen above
the western horizon in the evening.

In his copy Regiomontanus omitted the columns for the elongation
and its equation. The format of these double argument tables is basi-
cally the same as that which we find in the tables for 1321 compiled
by John of Murs, where the vertical argument is also mean anomaly,
given in time, and the horizontal argument is mean argument of center.
However, John of Murs's tables contain far fewer entries; for instance,
for each value of the horizontal argument there is only an entry in a
single column whereas Bianchini's tables have eight columns.

Table 41: True position of Mars (excerpt)

0,0°

Dies	Locus (°) Directus	Equatio (°) minue	ad di. (°/d) adde	ad ho. (°/h) adde	Lati. (°) Sept.	Equatio (°) minue	Elongatio (°) [Sun]	Equatio (°) adde
0	0; 0	0; 6,48	0;38,18	0; 1,36	0; 5	0; 0, 0	5,58;32	0; 5,18
10	6;23	0; 6,42	0;38,18	0; 1,36	0; 7	0; 0, 0	1;45	0; 5,30
20	12;46	0; 6,36	0;38,18	0; 1,36	0; 9	0; 0, 0	5; 2	0; 5,48
...								
770	37;14	0; 4,42	0;39,12	0; 1,38	0; 5	0; 0, 6	5,59;35	0; 4,30
779	43; 7	0; 4,30	[blank]	0; 1,38	0; 4	0; 0, 0	2;34	0; 4,30

0,10°

Dies	Locus (°) Directus	Equatio (°) minue	ad di. (°/d) adde	ad ho. (°/h) adde	Lati. (°) Sept.	Equatio (°) minue	Elongatio (°) [Sun]	Equatio (°) adde
0	5,58;52	0; 6,36	0;38,24	0; 1,36	0; 5	0; 0, 6	5,59;25	0; 5,24
10	5; 6	0; 6,18	0;38,24	0; 1,37	0; 7	0; 0, 0	2;40	0; 5,30
20	11;40	0; 6,12	0;38,36	0; 1,37	0; 9	0; 0, 0	6; 0	0; 5,42
...								
770	36;27	0; 4, 6	0;39,12	0; 1,39	0; 4	0; 0, 0	0;20	0; 4,18
779	42;22	0; 3,18	[blank]	0; 1,39	0; 4	0; 0, 0	3;19	0; 4, 6

Table 56: True position of Mercury (excerpt)

0,0°

Dies	Locus (°) Directus	Equatio (°) minue	ad di. (°/d) adde	ad ho. (°/h) adde	Lati. (°) meri.	Equatio (°) adde	Elongatio (°) [Mercury]	Equatio (°) minue
0	0; 0	0; 2, 6	1;42,20	0; 4,16	0;45	0; 1,42	1;56	0; 3,24
3	5;27	0; 1,54	1;43, 0	0; 4,17	1; 4	0; 1,42	4; 2	0; 3,18
6	10;16	0; 1,54	1;41,40	0; 4,14	1;25	0; 1,24	6;10	0; 3,24
...		_adde_					[Sun]	_adde_
112	1,44;52	0; 1, 0	1;53,20	0; 4,43	1;57	0; 0,24 _minue_	7;14	0; 0,18
115	1,50;32	0; 1,12	1;53, 0	0; 4,43	1;56	0; 0,18	4;34	0; 0, 6

Table 56 (*cont.*)

				0,10°			
Dies	*Locus* (°) *Directus*	*Equatio* (°) *minue*	*ad di.* (°/d) *adde*	*ad ho.* (°/h) *adde*	*Lati.* (°) *meri.*	*Equatio* (°) *adde*	*Elongatio* (°) [Mercury]	*Equatio* (°) *minue*
0	5,59;39	0; 1,54	1;43, 0	0; 4,18	0; 2	0; 1,30	1;22	0; 3,24
3	4;48	0; 1,42	1;43, 0	0; 4,17	1;21	0; 1,18	3;30	0; 3,24
6	9;57	0; 1,36	1;42,40	0; 4,17	1;39	0; 1,18	5;36	0; 3,24
...		*adde*					[Sun]	
112	1,45; 2	0; 1,18	1;54, 0	0; 4,45	2; 1	0; 0, 0 *minue*	7;17	0; 0,24
115	1,50;44	0; 1,18	1;54, 0	0; 4,45	1;53	0; 0,24	4;35	0; 0,36

In a period of anomaly the planet, initially at the apogee on its epicycle, returns to it and travels 6 physical signs; the intermediary tabulated values of $\lambda'(t - t_0, \bar{\kappa})$ represent the increment of the true longitude at the beginning of each of the days listed. This increment is not counted from Aries 0°; rather, it is an increment over the mean longitude at the time when the mean anomaly was 0°.

By inspection of the table we see that the three equations were derived in the same way according to the expressions:

$$q_L(t - t_0, \bar{\kappa}) = [\lambda'(t - t_0, \bar{\kappa}) - \lambda'(t - t_0, \bar{\kappa} + 10°)]/10,$$
$$q_B(t - t_0, \bar{\kappa}) = [\beta(t - t_0, \bar{\kappa}) - \beta(t - t_0, \bar{\kappa} + 10°)]/10,$$
$$q_E(t - t_0, \bar{\kappa}) = [\eta(t - t_0, \bar{\kappa} + 10°) - \eta(t - t_0, \bar{\kappa})]/10.$$

Similarly, the tabulated daily velocity results from the following expression:

$$v_D(t - t_0, \bar{\kappa}) = [\lambda'(t - t_0 + 10d, \bar{\kappa}) - \lambda'(t - t_0, \bar{\kappa})]/10,$$

whereas the tabulated hourly velocity results from:

$$v_H(t - t_0, \bar{\kappa}) = v_D(t - t_0, \bar{\kappa})/24.$$

The previous expression for v_D applies for all planets, except Mercury; since the entries for this planet are given at intervals of 3 days, the above expression turns into:

$$v_D(t - t_0, \bar{\kappa}) = [\lambda'(t - t_0 + 3d, \bar{\kappa}) - \lambda'(t - t_0, \bar{\kappa})]/3.$$

These quantities are used for interpolation purposes. Chapter 14 provides a succinct explanation of the use of the tables for the planets, and in Chapter 18 we are given a worked example for Mars for November 17, 1447 at 18;40h. Although not stated explicitly, the presentation of the double argument tables is that of a set of templates for computing the planet's true longitude at a given time. The general method to compute the true longitude, the latitude, and the true elongation of a planet by means of Bianchini's tables consists, first, in computing from the tables for mean motions the mean anomaly and the mean longitude of the planet at a given time. Then one has to find the time when, most recently, the mean anomaly was 0° and the argument of center corresponding to that time. Next, one enters the double argument table whose heading is closest to that mean argument of center and looks for the entries for the longitude, the latitude, and the true elongation closest to the time since the moment when the mean anomaly was, most recently, equal to 0° (i.e., subtracting the time corresponding to the number of completed cycles of anomaly). Then one interpolates in the double argument table to take into account the excesses of time and argument of center. The interpolation is greatly facilitated by the tabulated quantities. After these interpolations, the true longitude of the planet is found by adding the resulting quantity to the mean longitude of the planet when its mean anomaly was, most recently, equal to 0°. As an illustration of this method, let us consider the following quantities, t, $\bar{\alpha}(t)$, $\bar{\lambda}(t)$, t_0, $\bar{\alpha}(t_0)$, $\bar{\kappa}(t_0)$, where t is the given time, and t_0 is the time when the anomaly was most recently equal to 0°. From the mean motion tables one finds $\bar{\alpha}(t)$ and $\bar{\lambda}(t)$. Then we seek t_0: for this we have to subtract from t the time corresponding to a number of complete cycles of the planet's anomaly such that $\bar{\alpha}(t - t_0)$ is the excess of mean anomaly since the time, t_0, when the mean anomaly was most recently 0°. In other words, $t - t_0$ cannot exceed the period of the planet's anomaly. To compute the mean argument of center for t_0, $\bar{\kappa}(t_0)$, we subtract the longitude of the apogee from $\bar{\lambda}(t_0)$. We then enter the double argument table with $t - t_0$ and $\bar{\kappa}(t_0)$. After suitable interpolations, we find the quantity λ', and the true longitude at t, $\lambda(t)$, is $\bar{\lambda}(t_0) + \lambda'(t - t_0, \bar{\kappa}(t_0))$. The quantity, λ', is the sum of the mean argument of center in the heading (i.e., at time t_0), the increment in the argument of center from t_0 to t, and the appropriate corrections for anomaly and argument of center. Thus

$$\lambda(t) = \bar{\lambda}(t_0) + \lambda' = \bar{\lambda}(t_0) + \bar{\kappa}(t_0) + \Delta\bar{\kappa}(t - t_0) + c(\bar{\alpha}, \bar{\kappa}),$$

where

$$\Delta\bar{\kappa}(t - t_0) = \bar{\kappa}(t) - \bar{\kappa}(t_0).$$

Then

$$\bar{\kappa}(t) = \bar{\kappa}(t_0) + \Delta\bar{\kappa}(t - t_0),$$

and

$$c(\bar{\alpha}, \bar{\kappa}) = c(\bar{\alpha}(t), \bar{\kappa}) = c(\bar{\alpha}(t - t_0), \bar{\kappa}).$$

In the worked example for Mars, we are first told to find in Tables 32 and 33 the entries for 1440y and 6y for the two quantities: time (*superatio*) and increment in longitude (*motus*). We are also told to add the number of days in October and 17d 18;40h to the time. Again, we note that Bianchini uses complete days. After adding the corresponding radices for the Incarnation found in Table 37, the resulting values are:

> time = 1751d 17;47h = 285d 14;12h (1440y) + 631d 3;15h (6y) + 304d (October) + 17d 18;40h + 513d 5;40h (Incar.)
> increment in longitude = 5,18;59° = 1,29;5° (1440y) + 1,37;29° (6y) + 2,12;25° (Incar.).

We are then told to subtract from the time shown above the entry corresponding to 2 periods of anomaly in Table 38, and to add to the increment in longitude the corresponding entry for 2 periods of anomaly. Again, we note that the *modus operandi* in Bianchini's tables, in order to use the tables as templates, is: subtract the periods in time, and add the corresponding entries for the other quantities. The resulting values are:

> time = 191d 21;1h = 1751d 17;47h – 1559d 20,46h (2 periods of anom.)
> increment in longitude = 56;27° = 5,18;59° – 1,37;28° (2 periods of anom.).

This means that 191d 21;1h before the time in question (Nov. 17 at 18;40h) the mean anomaly was 0°, and the mean longitude of the planet, $\bar{\lambda}_0 = 56;27°$. This is exactly what follows from recomputing the longitude of Mars for May 9, 1447, a date for which the recomputed mean anomaly is 0°.

Then we must subtract the apogee of Mars for that time (2,14;31°) from the increment of longitude, shown above, to obtain 4,41;56°. The text refers to this quantity as *centrum*. We note that $\bar{\kappa}$, the argument of center, is 4,41;56° when the mean anomaly, $\bar{\alpha}$, is 0°, not that of November

17, 1447. We are then instructed to enter Table 41 with $t - t_0 = 190d$ and $\bar{\kappa} = 4{,}40° = 280°$, the closest values to the computed data found in the table. The corresponding entries in the double argument table are:

$\lambda'(190, 280) = 2{,}7{;}25°$
$q_L(190, 280) = -0{;}7{,}12°$
$v_D(190, 280) = 0{;}34{,}36°/d$
$v_H(190, 280) = 0{;}1{,}26°/h$
$\beta(190, 280) = +0{;}51°$
$q_B(190, 280) = -0{;}0{,}6°$
$\eta(190, 280) = 58{;}46°$
$q_E(190, 280) = +0{;}9{,}18°.$

Next come rules for interpolation: we are told to consider the excess of argument of center over 4s 40° ($1{;}56° \approx 2°$) and to multiply it by $q_L(190, 280)$, $q_B(190, 280)$, and $q_E(190, 280)$, to obtain the corresponding corrections to the other quantities:

$\Delta'\lambda = -0{;}14° = -0{;}7{,}12° \cdot 2$
$\Delta'\beta = 0° = -0{;}0{,}6° \cdot 2$
$\Delta'\eta = 0{;}18° = +0{;}9{,}18° \cdot 2,$

where $\Delta'\lambda$, $\Delta'\beta$, and $\Delta'\eta$, are the excesses of longitude, latitude, and elongation, respectively. These corrections have to be added to $\lambda'(190, 280)$, $\beta(190, 280)$, and $\eta(190, 280)$, respectively. We also have to consider the excesses of time over 190d, in days (1d) and hours (21;1h), and to multiply them by $v_D(190, 280)$ and $v_H(190, 280)$, respectively, to obtain:

correction due to the excess in days $= 0{;}34{,}36° = 0{;}34{,}36°/d \cdot 1d$
correction due to the excess in hours $= 0{;}30° = 0{;}1{,}26°/h \cdot 21{;}1h.$

These two corrections add up to a correction due to the excess in time, $c_i(\Delta't) = 1{;}5°$, and it has to be added to $\lambda'(190, 280)$, $\beta(190, 280)$, and $\eta(190, 280)$, respectively. Finally, the true longitude of Mars (on the 9th sphere) results from adding $\lambda'(190, 280)$, the excess of longitude ($-0{;}14°$), and the correction due to the excess in time ($1{;}5°$) to the longitude of the planet when the mean anomaly was 0° ($56{;}27°$):

$3{,}4{;}43° = 56{;}27° + 2{,}7{;}25° - 0{;}14° + 1{;}5°,$

or, in general,

$\lambda = \bar{\lambda}_0 + \lambda'(t - t_0, \bar{\kappa}) + \Delta'\lambda + c_i(\Delta't),$

where $t - t_0$ is the argument in the rows to the nearest day less than the time deduced from the time in question and the time when the mean anomaly was most recently 0°, and $\bar{\kappa}$ is the argument of center for the columns which is the nearest multiple of 10° less than the argument of center at the time when the anomaly was most recently 0°.

Similarly, we find the true latitude of Mars by interpolating from $\beta(190, 280)$ both in time (mean anomaly) and in argument of center. The excess in latitude due to the excess of argument of center is 0°, as mentioned above, and the excess in latitude due to the excess of time (1d 21;1h), $c_2(\Delta't)$, is also 0°, according to Bianchini (the increment in latitude between 190d and 200d amounts to $0;6° = 0;57° - 0;51°$). Hence, the true latitude is

$$\beta(190, 280) + 0° = +0;51°$$

or, in general,

$$\beta = \beta(t - t_0, \bar{\kappa}) + \Delta'\beta + c_2(\Delta't),$$

where $t - t_0$ and $\bar{\kappa}$ are defined as before.

In order to find the true elongation, that is, the difference between the true longitude of the Sun and that of the planet, we are told to find the excess of elongation due to the excess of time (1d 21;1h), which according to Bianchini is $c_3(\Delta't) = 0;50,9°$ (the increment in elongation between 190d and 200d amounts to $4;27° = 1,3;13° - 58;46°$), and to add it to the excess of elongation due to the argument of center, $\Delta'\eta = 0;18°$, as mentioned above. Then, the true elongation of Mars at the given time results from adding these two excesses to $\eta(190, 280)$, and the result is

$$59;54,9° (= 58;46° + 0;50,9° + 0;18°)$$

or, in general,

$$\eta = \eta(t - t_0, \bar{\kappa}) + \Delta'\eta + c_3(\Delta't),$$

where $t - t_0$ and $\bar{\kappa}$ are defined as before.

Using the standard Alfonsine Tables, we have recomputed the entries for $t - t_0 = 190d$ and $\bar{\kappa} = 4,40° = 280°$ in this table, namely, $\lambda'(190, 280) = 2,7;25°$, $\beta(190, 280) = +0;51°$, and $\eta(190, 280) = 58;46°$. In 190 days the increment in longitude is

$$\Delta\bar{\lambda} = 99;34° (= 190d \cdot 0;31,26,39°/d)$$

and we define $\bar{\kappa}'$ such that:

$\overline{\kappa}' = \overline{\kappa} + \Delta\overline{\lambda} = 280° + 99;34° = 19;34°.$

Also, in 190 days, the increment in mean anomaly is

$\overline{\alpha} = 87;42°$ (= 190d · 0;27,41,41°/d).

The equation of center, $c_3(\overline{\kappa}')$, corresponding to 19;34°, is −3;30°, and the value for the minutes of proportion, $c_4(\overline{\kappa}')$, is 56. Since, in this table, t − t_0 (190d in our case) is the time elapsed after the instant when the mean anomaly was 0°, the true anomaly, $\alpha = \overline{\alpha} + c_3(\overline{\kappa}')$, is 91;12° (= 87;42° + 3;30°). The equation of argument, $c_6(\alpha)$, corresponding to 91;12°, is +33;44°, and the correction, $c_5(\alpha)$, is −2;30° (where c_i refers to a value computed with interpolation from the entries in column i in the standard Alfonsine Tables). According to the rules for computing a planetary position with the Alfonsine Tables,

$$\lambda' = (t- t_0, \overline{\kappa}) = \Delta\overline{\lambda} + c_3(\overline{\kappa}') + c_6(\alpha) + c_4(\overline{\kappa}') \cdot c_5(\alpha).$$

Hence,

$$\lambda'(190, 280) = 99;34° - 3;30° + 33;44° - (2;30° \cdot 0;56) = 127;18°,$$

in close agreement with the given entry (127;25°).

In order to find the true longitude of Mars at t, one must add λ' to the mean longitude of Mars at the time when the mean anomaly is 0°, most recently before t.

In Figure 3, the observer is at O, the center of the deferent is at D, the equant point is at E, the center of the epicycle is at C, the vernal point is in the direction OV, the mean position of Mars is in the direction $\overline{OM}$, the mean Sun is in the direction $\overline{OS}$, the mean epicyclic apogee is at $\overline{A}_e$, the true epicyclic apogee is at A_e, $\overline{OS}$ is parallel to CM, and the true position of Mars is at M. There are two equations: q_1 is approximated by $c_3(\overline{\kappa}')$, and q_2 is approximated by $c_6(\alpha) + c_4(\overline{\kappa}') \cdot c_5(\alpha)$.

The entry for the latitude, $\beta(190, 280) = +0;51°$, results from multiplying the northern latitude (a function of true anomaly) by the minutes of proportion (a function of the true center). The true anomaly at that time computed above was 91;12°. To obtain the true center it is necessary to correct the previously found mean center, $\overline{\kappa}' = 19;34°$, by the amount of the corresponding equation of center, −3;30°; hence, the true center is 16;4° (= 19;34° − 3;30°). In the tables for the latitude of Mars (Table 73) we find +0;53° for the northern latitude and 57;31 for the minutes of proportion, after interpolation. Their product is +0;50,48° ≈ +0;51°, in agreement with the text.

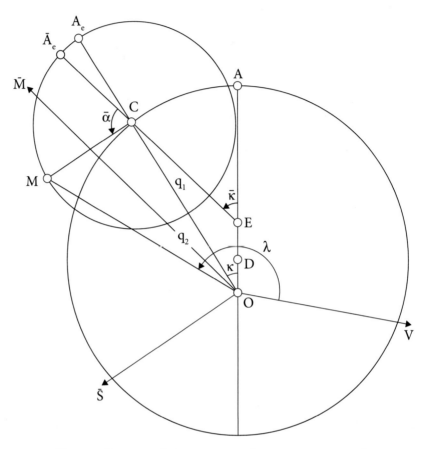

Figure 3: Position of Mars on November 17, 1447 at 18;40h.

Finally, to recompute the entry for elongation, $\eta(190, 280) = 58;46°$, one needs to know the true longitudes of the Sun and Mars at time t. As stated above, the true longitude of Mars is the sum of $\lambda'(190, 280) = 127;25°$ and its mean longitude when $\bar{\alpha}$ is $0°$ ($56;27°$). The result is $183;52°$. As for the true longitude of the Sun at time t, one must first determine the mean solar longitude at time t. This results from adding the mean solar anomaly found above ($87;42°$) to the mean longitude of Mars at time t ($156;1,23° = 56;27° + 190d \cdot 0;31,26,39°/d$), where $0;31,26,39°/d$ is the mean motion in longitude of Mars. The result is $243;43°$. Hence, the mean solar anomaly at time t is $243;43° - 90;45° = 152;58°$, where $90;45°$ is the longitude of the solar apogee at that time. The solar equation corresponding to $152;58°$ is $-1;0,43° \approx -1;1°$, and thus the true solar longitude at time t is 242;42° (= 243;43° − 1;1°). Finally, η(190, 280) = 242;42° − 183;52° = 58;50°, in agreement with the given entry.

Table 57. Mean motions for conjunction of the luminaries in collected years

Na, f. 94v: *Tabula* (…) *luminarium*
Rc, f. 47r: *Tabula radicum coniunctionium luminarium*
Ed. 1495, f. C6r–v: *Tabula radicum coniunctionis luminarium*
Ed. 1526, f. 230r–v: *Tabula radicum* [symbol for conjunction] *luminarium*

In this table there are five columns: see Table 57. The first is for the argument: collected years from 40 to 2000 at intervals of 40y. Thus, all the entries correspond to leap years. The entries in the other columns represent the time (in days, hours, and minutes), the longitude of the luminaries at mean conjunction (in physical signs, degrees, and minutes), the mean lunar anomaly (in degrees and minutes), and the argument of lunar latitude (in physical signs, degrees, and minutes). The time displayed in column 2, headed *superatio*, is the excess over an integer number of mean synodic months; hence, the entries in this column do not exceed 29d 12;44h, which is the length of the mean synodic month used here. This value is mentioned in Chapter 22 and is explicitly given in Table 59, below. However, the entries in these tables were computed with a higher precision. Consider, for instance, the entry for 2000 years, 1d 18;35h. In 2000 years there are 2000y · 365;15d = 730,500 days and in 730,498d 5;25h (= 730,500d − 1d 18;35h) there is an integer number of mean synodic months, which happens to be 24,737. Dividing the time interval corresponding to 24,737 mean synodic months by that number, we find a length of the mean synodic month of 29d 12;44,3,3h, which is exactly the value generally used in Alfonsine astronomy. As was the case for previous tables, the entries for the solar longitude at conjunction, the lunar anomaly, and the argument for lunar latitude do not correspond to the exact number of years indicated in the first column; rather, they correspond to the nearest previous conjunction of the Sun and the Moon. This table presents data that are analogous to what we find in Table 60, and in the comments to that table we offer some sample computations of its entries.

Table 57: Mean motions for conjunction of the luminaries in collected years
(excerpt)

Years	Time (d, h)		Longitude (°)	Anomaly (°)	Arg. latitude (°)
40	21	21;19*	5,38;43	153;31	31;13
80	14	5;54	5,46;33	332;51	1,33; 7
...					
1400	27	19;40	5,42;52	258;23	59;33
1440	20	4;15	5,50;42	77;43	2, 1;27
...					
2000	1	18;35	12;17	350;53	2,55;57

* With MS Na. An erroneous reading, 21d 21;16h, is found in both ed. 1495 and ed. 1526.

Table 58. Radices for conjunction of the luminaries

Na, f. 94v: *Radices*
Rc, f. 47r: *Tabula radicum*
Ed. 1495, f. C6v: *Tabula radicum*
Ed. 1526, f. 230v: *Tabula radicum*

The five columns in this table are the same as in Table 57: see Table 58. Ed. 1526 adds entries for 1480.

Table 58: Radices of the Moon for conjunction

	Time (d, h)		Longitude (°)	Anomaly (°)	Arg. latitude (°)
Incarnation	16	17; 4	4,21;49	339;55	5,52;52
1400	14	23;59	4,33;48	264; 7	1,23; 6
1440	7	8;34	4,41;37	83;27	2,24;59

The entry 16d 17;4h for the Incarnation tells us that a mean conjunction of the Sun and the Moon took place 16d 17;4h before the Incarnation. In 16d 17;4h the Moon progresses in longitude 220;11,32° at the standard mean motion, 13;10,35°/d. The longitude of the Moon for Ferrara at the Incarnation was 2,2;0,43° = 122;0,43°;[15] 16d 17;4h earlier it was 4,21;49°

---

[15] See Ratdolt, *Tabule astronomice Alfontij*, d1v; see also Poulle, *Les tables alphonsines*, 127.

(= 122;0,43° – 220;11,32°), in agreement with the entry. Similarly for the anomaly: in 16d 17;4h the Moon progresses in anomaly 218;19,50° at the standard mean motion, 13;3,54°/d. The lunar anomaly for the Incarnation and the longitude of Ferrara was 3,18;14,31° = 198;14,31°;[16] hence, 16d 17;4h earlier it was 339;54,41° (=198;14,31° – 218;19,50°). As for the argument of lunar latitude: in 16d 17;4h the Moon progresses 221;4,43° at the standard mean motion, 13;13,46°/d, and the argument of lunar latitude for Ferrara at the Incarnation was 3,33;42,24° = 213;42,24°.[17] Thus, the argument of lunar latitude 16d 17;4h before the Incarnation was 5,52;38° (= 213;42,24° – 221;4,43°). Again, as occurred in Table 15, this result differs from the entry in Table 58 by 0;14°.

As was the case for the radices of the Sun, the lunar node, and the Moon for anomaly, the entries for the radices in 1400 and 1440 in Table 58 are obtained by adding those for the Incarnation to those corresponding to 1400 and 1440 years in the tables of the mean motions. Thus, for 1400 we obtain:

> 14d 24;0h = 16d 17;4h (Incarn.) + 27d 19;40h (1400y) – 29d 12;44h (1 syn. month),
> 4,33;47° = 4,21;49° (Incarn.) + 5,42;52° (1400y) + 29;6° (1 syn. month),
> 264;7° = 339;55° (Incarn.) + 258;23° (1400y) + 25;49° (1 syn. month),
> 1,23;5° = 5,52;52° (Incarn.) + 59;33° (1000y) + 30;40° (1 syn. month),

in close agreement with the tabulated values.

Table 59. Multiples of half a mean synodic month

Na, f. 94v: *Tabula brevis*
Rc, f. 47r: *Tabula brevis revolutionum*
Ed. 1495, f. C6v: *Tabella brevis revolutionum*
Ed. 1526, f. 230v: *Tabella brevis revolutionum*

This table has the same presentation as Table 57, but there are two columns for the longitude of the luminaries at mean conjunction, one for the Sun and another for the Moon. At the far right there is a symbol,

---

[16] See Ratdolt, *Tabule astronomice Alfontij*, d1v; see also Poulle, *Les tables alphonsines*, 128.
[17] See Ratdolt, *Tabule astronomice Alfontij*, d1v; see also Poulle, *Les tables alphonsines*, 128.

alternately for opposition, O, and conjunction, C. In this case we are given entries for the first eight multiples of half a mean synodic month: see Table 59.

Table 59: Multiples of half a mean synodic month (excerpt)

Days	(h)	Solar long. (°)	Lunar long. (°)	Anomaly (°)	Arg. latitude (°)	
14	18;22	14;33	3,14;33	192;55	3,15;20	O
29	12;44	29; 6	29; 6	25;49	30;40	C
...						
103	8;34	1,41;52	4,41;52	270;22	4,47;21	O
118	2;56	1,56;26	1,56;26	103;16	2, 2;48	C

The entries for 29d 12;44h, the length of the mean synodic month, are derived from the standard parameters, for

$$29;6° = 29d\ 12;44h \cdot 0;59,8°/d,$$
$$25;49° = (29d\ 12;44h \cdot 13;3,54°/d) - 360°,$$
$$30;40° = (29d\ 12;44h \cdot 13;13,46°/d) - 360°,$$

where 0;59,8°/d, 13;3,54°/d, and 13;13,46°/d are the mean motions in solar longitude, lunar anomaly, and argument of lunar latitude, respectively.

Table 60. Mean motions for conjunction of the luminaries in expanded years

Na, f. 94v: *Tabula (...) luminarium*
Rc, f. 47r: *Tabula radicum coniunctionium luminarium*
Ed. 1495, f. C7r: *Tabula radicum coniunctionis luminarium*
Ed. 1526, f. 231r–v: *Tabula radicum coniunctionis luminarium*

The layout of this table is the same as that of Table 57, but here the argument is the number of expanded years, from 1y to 40y: see Table 60. The entries for 40 expanded years agree with those for 40 collected years in Table 57.

Table 60: Mean motions for conjunction of the luminaries in expanded years (excerpt)

Years	Time (d, h)		Longitude (°)	Anomaly (°)	Arg. latitude (°)
1	10	15;11	5,49;17	309;48	8; 3
2	21	6;23	5,38;34	259;36	16; 6
...					
39	10	6; 7	5,49;26	203;43	23;11
40	21	21;19	5,38;43	153;31	31;13

The entry for one year, 10d 15;11h, indicates that in a year of 365 days there is an integer number of mean synodic months plus 10d 15;11h and, indeed,

10d 15;11h = 365d − (12 · 29d 12;44h).

Then, the entry 5,49;17° for the longitude at conjunction (1 year) has to be understood as the increment in longitude between the beginning of the year and this precise moment (10d 15;11h before the completion of the year), that is, the increment in longitude in 12 synodic months. Indeed,

5,49;17° = (365d − 10d 15;11h) · 0;59,8,20°/d,

and

5,49;17° = 12 · 360° − [(365d − 10d 15;11h) · 13;10,35,1°/d].

The entries for one year for the lunar anomaly, 309;48°, and for the argument of lunar latitude, 8;3°, are obtained analogously:

309;48° = [(365d − 10d 15;11h) · 13;3,53,57°/d] − 12 · 360°,
8;3° = [(365d − 10d 15;11h) · 13;13,45,39°/d] − 13 · 360°.

Table 61. Monthly mean motions for conjunction of the luminaries

Na, f. 94v: *Menses non bisextiles* and *Menses bisextiles*
Rc, f. 47r: *Menses non bisextiles* and *Menses bisextiles*
Ed. 1495, f. C7v: *Menses non bisextiles* and *Menses bisextiles*
Ed. 1526, f. 231v: *Menses non bisextiles* and *Menses bisextiles*

This table has two sub-tables, one for a common year and another for a leap year, both beginning in January: see Table 61. The layout of this table is the same as that of Table 57, but here the first column lists the names of the 12 months. Table 61 shows some entries of the first sub-table. The entry for January (time) results from subtracting the mean synodic month (29d 12;44h) from 31 days, and the entries for the three other quantities correspond to one synodic month (see Table 59, line 2). The entries for December are 10d 15;11h (time) in a common year and 11d 15;11h (time) in a leap year. The other three entries for December are the same both for a common and a leap year. We note that all entries for December in a common year agree with those for 1 year in the preceding table.

Table 61: First sub-table for the monthly mean motions for conjunction in a common year (excerpt)

	Common year (d, h)		Solar long. (°)	Anomaly (°)	Arg. latitude (°)
January	1	11;16	0,29; 6	25;49	30;40
February	29	11;16	0,29; 6	25;49	30;40
March	1	9;48	1,27;19	77;27	1,32; 1
...					
November	9	3;55	5,20;10	283;59	5,37;23
December	10	15;11	5,49;17	309;48	8; 3

Table 62. Time of mean conjunctions during the year

Na, f. 107r–v: *Tabula temporis coniunctionum in annis bisextilibus* and *Tabula (…) anno bisextili*

Rc, f. 47v: *Tabula temporis coniunctionum mediarum luminarium in annis non bisextilibus* and *Tabula temporis coniunctionum mediarum luminarium in annis bisextilibus*

Ed. 1495, ff. C8r–D1v: *Tabula* [symbol for conjunction] *in annis non bissextilibus* and *Tabula* [symbol for conjunction] *in anno bissextili*

Ed. 1526, ff. 232r–233v: *Tabula* [symbol for conjunction] *in annis non bisextilibus* and *Tabula* [symbol for conjunction] *in anno bisextili*

This table, presented as two sub-tables, one for a common year and another for a leap year, gives the amount of time to be added to a mean conjunction in January—whose time is known—in order to obtain the times of the 12 following conjunctions: see Table 62. It is a different and more expanded way to present the information contained in Table 59, and has the advantage that we are also given the day of the month on which each of the conjunctions takes place. The use of this table is explained in Chapter 22, without a worked example.

Table 62: Time of mean conjunctions for a common year (excerpt)

January (d, h)	February (d, h)	March (d, h)	...	November (d, h)	December (d, h)	January (d, h)
0 0; 0		0 1;28				
29 12;44	0 0; 0	29 14;12	...	20 20; 5	20 8;49	18 21;33,37
1 0; 0		1 1;28				
30 12;44	0 0; 0	30 14;12	...	21 20; 5	21 8;49	19 21;33,37
2 0; 0		2 1;28				
31 12;44	0 0; 0	31 14;12	...	22 20; 5	22 8;49	20 21;33,37
3 0; 0	1 12;44	3 1;28				
4 0; 0	2 12;44	4 1;28				
...						
30 0; 0	28 12;44	30 1;28	...	20 7;21	19 20; 5	18 8;49,37
	1 12;44	1 12;44	...	21 7;21	20 20; 5	19 8;49,37
31 0; 0	29 12;44	31 1;28	...	22 7;21	21 20; 5	20 8;49,37

Table 63. True position of the Sun at mean conjunction

Na, ff. 87r–94r: *Tabula solis in coniunctionibus*
Rc, ff. 49r–54v: *Tabula solis in coniunctionibus*
Ed. 1495, ff. D2r–F8r: no title
Ed. 1526, ff. 234r–256r: no title

In this double argument table, the vertical argument (not specified) is the number, from 0 to 12, corresponding to each mean synodic month in a year for a total of 13 rows, and the horizontal argument is the mean solar anomaly, from 0,0° to 5,59°, at 1°-intervals. For each conjunction, numbered [C], and each value of the mean solar anomaly, $\bar{\kappa}$, we are given two entries: the increment in true solar longitude (in physical signs, degrees, and minutes), and the hourly velocity of the Sun (in minutes and seconds): see Table 63.

Table 63: True position of the Sun at mean conjunction (excerpt)

| [C] | 0,0° | | 0,1° | |
	Locus (°)	ad. ho. (°/h)	Locus (°)	ad ho. (°/h)
[0]	0; 0	0; 2,23	5,59;58	0; 2,23
[1]	28; 5	0; 2,23	28; 3	0; 2,23
[2]	56;25	0; 2,25	56;23	0; 2,25
[3]	1,25;10	0; 2,27	1,25;10	0; 2,27
[4]	1,54;28	0; 2,30	1,54;29	0; 2,30
[5]	2,24;16	0; 2,33	2,24;18	0; 2,33
[6]	2,54;25	0; 2,34	2,54;26	0; 2,34
[7]	3,24;39	0; 2,33	3,24;42	0; 2,33
[8]	3,54;37	0; 2,31	3,54;39	0; 2,31
[9]	4,24; 7	0; 2,28	4,24; 7	0; 2,28
[10]	4,53; 4	0; 2,26	4,53; 3	0; 2,26
[11]	5,21;31	0; 2,24	5,21;29	0; 2,24
[12]	5,49;40	0; 2,23	5,49;38	0; 2,23

Let $\lambda_S'([C], \overline{\kappa})$ and $v_S([C], \overline{\kappa})$ be the values tabulated in the two columns, respectively. In 12 consecutive mean synodic months of 29d 12;44h the Sun travels almost 6 physical signs; the intermediary tabulated values of $\lambda_S'([C], \overline{\kappa})$ represent the increment of the true solar longitude at the beginning of each of the 13 months listed. This increment is counted from the conjunction in row 0, not from Aries 0°, that is, $\overline{\kappa}$ in the heading refers to the first row in each sub-table. Thus, for example, to compute $\lambda_S'([1], 0)$, one needs to know the increment in mean solar longitude in one synodic month, 29;6° (see Table 59). The corresponding solar equation is –1;1° and, then,

$$\lambda_S'([1], 0) = 28;5° = 29;6° - 1;1° \text{ (text: 28;5°).}$$

Similarly, to compute $\lambda_S'([0], 1)$, one needs to know the solar equation corresponding to a mean solar anomaly of 1°. The equation is –0;2,10°, and the true solar anomaly is 0;57,50° (= 1;0° – 0;2,10°). Then,

$$\lambda_S'([0], 1) = 5,59;58° = 6,0° + 0;57,50° - 1;0° \text{ (text: 5,59;58°).}$$

In a general case such as $\lambda_S'([4], 1)$, the mean longitude in four mean synodic months has increased by 1,56;26° (see Table 59). The mean solar anomaly, i.e., the sum of the value in the heading (1°) and the increment (1;56;26°), is thus 1,57;26°. The corresponding solar equation is –1;57°, and the true solar anomaly is 1,55;29° (= 1,57;26° – 1;57°). Then

$\lambda_S'([4], 1) = 1,54;29° = 1,55;29° - 1°$ (text: $1,54;29°$).

The tabulated hourly velocity may be taken from the first row of Table 63 ([C] = 0) in the sub-table for the appropriate mean solar anomaly.

Table 64. True position of the Moon at mean conjunction

Na, ff. 95r–106v: *Tabula lune in coniunctionibus*
Rc, ff. 55r–63v: *Tabula lune in coniunctionibus*
Ed. 1495, ff. F8v–M5r: *Tabula lune in coniunctionibus*
Ed. 1526, ff. 257r–301r: *Tabula lune in coniunctionibus*

In this double argument table, the vertical argument (not specified) is the number from 0 to 12 corresponding to each mean synodic month in a year, as in the previous table, and the horizontal argument is the mean lunar anomaly, from 1° to 360°, at 1°-intervals. For each conjunction, numbered [C], and each value of the mean lunar anomaly, $\overline{\alpha}$, we are given three entries: the increment in true lunar longitude (in physical signs, degrees, and minutes), the hourly velocity of the Moon (in minutes and seconds), and the true argument of lunar latitude (in physical signs, degrees, and minutes): see Table 64.

Table 64: True position of the Moon at mean conjunction (excerpt)

[C]	1°			2°		
	Locus (°)	*ad. ho.* (°/h)	*Arg. lati.* (°)	*Locus* (°)	*ad ho.* (°/h)	*Arg. lati.* (°)
[0]	5,59;55	0;29,37	5,59;55	5,59;50	0;29,37	5,59;50
[1]	27; 1	0;29,58	28;35	26;57	0;29,59	28;31
[2]	54;29	0;30,50	57;37	54;26	0;30,53	57;34
[3]	1,22;36	0;32, 8	1,27;17	1,22;34	0;32,12	1,27;15
[4]	1,51;34	0;33,45	1,57;49	1,51;35	0;33,49	1,57;50
[5]	2,21;33	0;35,22	2,29;22	2,21;36	0;35,25	2,29;25
[6]	2,52;28	0;36,33	3, 1;51	2,52;33	0;36,35	3, 1;56
[7]	3,23;56	0;36,51	3,34;53	3,24; 2	0;36,50	3,34;59
[8]	3,55;21	0;36, 9	4, 7;52	3,55;26	0;36, 6	4, 7;57
[9]	4,26; 6	0;34,47	4,40;10	4,26; 9	0;34,43	4,40;13
[10]	4,55;58	0;33, 6	5,11;36	4,55;59	0;33, 2	5,11;37
[11]	5,24;49	0;31,34	5,42; 1	5,24;47	0;31,31	5,42;59
[12]	5,52;48	0;30,25	11;34	5,52;44	0;30,23	11;30

Let $\lambda_L'([C], \overline{\alpha})$, $v_L([C], \overline{\alpha})$, and $\omega([C], \overline{\alpha})$ be the values tabulated in the three columns, respectively. In 12 consecutive mean synodic months of 29d 12;44h the Moon travels almost 6 physical signs in longitude; the intermediary tabulated values of $\lambda_L'(C, \overline{\alpha})$ represent the increment of the true solar longitude at the beginning of each of the 13 months listed. As was the case for the Sun, this increment is counted from the conjunction in row 0, not from Aries 0°, that is, $\overline{\alpha}$ in the heading refers to the first row in each sub-table. Thus, for example, in order to compute $\lambda_L'([0], 1)$, we need to know the equation of argument for $\overline{\alpha}$ = 1°, which is $c_6(\overline{\alpha}) = -0;5°$; hence,

$$\lambda_L'([0], 1) = 360° - 0;5° = 359;55°,$$

in agreement with the entry in the text. To compute $\lambda_L'([1],1)$, we have to take into account that the increments in mean lunar longitude and mean lunar anomaly in one synodic month are 29;6° and 25;49°, respectively (see Table 59). In this case, the equation of argument is $c_6(\overline{\alpha} + 1°) = c_6(26;49°) = -2;3°$; hence,

$$\lambda_L'([1], 1) = 29;6° - 2;3° = 27;3° \text{ (text: 27;1°)}.$$

Similarly, to compute $\lambda_L'([0], 2)$, one needs to know the equation of argument corresponding to a mean anomaly of 2°. The equation is $-0;9,31°$, and the increment of true longitude is

$$\lambda_L'([0], 2) = 5,59;50° = 6,0° - 0;9,31° \text{ (text: 5,59;50°)}.$$

In a general case such as $\lambda_L'([4], 1)$, in four mean synodic months the mean longitude has increased by 1,56;26° and the mean anomaly by 103;16° (see Table 59). The mean lunar anomaly, that is, the sum of the value in the heading (1°) and the increment (103;16°), is thus 104;16°. The corresponding equation of argument is $c_6(104;16°) = -4;55°$; hence, the increment in true longitude is

$$\lambda_L'([4], 1) = 1,56;26° - 4;55° = 1,51;31° \text{ (text: 1,51;34°)}.$$

As in Table 17, the true argument of latitude, $\omega([C], \alpha)$, is related to the increment of true longitude by means of the expression:

$$\omega([C], \alpha) = \lambda_L'([C], \alpha) - \lambda_N,$$

where λ_N is the longitude of the lunar node which, in Alfonsine astronomy, moves at the rate of $-0;3,10,38°$/d. Indeed, this is the quantity embedded in the table such that, for any ω and λ_L', the following relation holds:

$$\omega([C], \alpha) = \lambda_L'([C], \alpha) - C \cdot 1;34°,$$

where $-1;34° = -0;3,10,38°/d \cdot 29d\ 12;44h$ is the distance traveled by the node in one mean synodic month.

Chapter 21 gives an example of the use of the tables for conjunctions. We are asked to compute the true conjunction that took place in August 1447.

Although not stated explicitly, the general method for computing the time of the true conjunction, the true solar longitude, the true lunar anomaly, and the true argument of lunar latitude, at that time by means of Bianchini's tables consists, first, in computing from the tables for mean motions the time of the mean conjunction, and the other three mean values at that time. Second, one has to find the mean solar anomaly. Third, one enters the tables for the true positions for the Sun and the Moon at conjunction and looks for the true longitudes and the argument of lunar latitude closest to the given mean solar and lunar anomalies. And, fourth, one interpolates in these tables to take into account the excesses of solar and lunar anomalies. In the worked example, we are first told to find in Tables 57, 60, and 61, the entries for 1440y, 6y, and July for the four quantities: time, mean longitude of the luminaries at mean conjunction, mean lunar anomaly, and mean argument of lunar latitude. We are also told to subtract the time from twice the length of the mean synodic month (59d 1;28h). After adding the corresponding radices for the Incarnation found in Table 58, the resulting values are:

time = 11d 3;37h = 59d 1;28h (2 syn. months) – [20d 4;15h (1440y) + 5d 17;40h (6d) + 5d 6;52h (July) + 16d 17;4h (Incar.)]
mean longitude: $\bar{\lambda}$ = 2,28;23° = 5,50;42° (1440y) + 5,53;54° (6d) + 3,23;45° (July) + 4,21;49° (Incar.) + 58;13° (2 syn. months)
mean lunar anomaly: $\bar{\alpha}$ = 40;26° = 77;43° (1440y) + 110;27° (6d) + 180;43° (July) + 339;55° (Incar.) + 51;38° (2 syn. months)
mean argument of latitude: $\bar{\omega}$ = 2,19;58° = 2,1;27° (1440y) + 1,49;37° (6d) + 3,34;42° (July) + 5,52;52° (Incar.) + 1,1;20° (2 syn. months).

The time computed above, 11d 3;37h, is the time in August in year 1447 when mean conjunction took place. In order to perform these computations, it should be kept in mind that for Tables 57, 60, and 61, which give the mean motions at conjunction, the epoch is the mean conjunction prior to the Incarnation, and thus the entries in these tables give the excess in time over a certain number of mean synodic months, each one of them not to exceed the length of a mean synodic month. Hence, the accumulated excess has to be subtracted from a multiple of a mean synodic month in order to get the time from the beginning of the month in which the mean conjunction occurs. As for the quantities other than time, it should be stressed that subtracting the accumulated excess from a multiple of the mean synodic month does not mean that the relevant time interval has not elapsed. Rather, the computation for the time is related to the calendar, for one wishes to determine the date and time of day of the mean conjunction in a specific month in the Julian calendar. The other quantities depend on the time of mean conjunction, but not on the calendar. Therefore, the increments in the other quantities corresponding to a multiple of mean synodic months have to be added to the accumulated increments for the given date.

We are then told to subtract the solar apogee at the given time, 1,30;40°, to find the mean solar anomaly, 57;43° (= 2,28;23° − 1,30;40°). This serves as the argument in Table 63 (first row), where we are given the entries:

$\lambda_S'([0], 58) = 5,58;11°$
$v_S([0],58) = 0;2,25°/h.$

The true longitude of the Sun, 2,26;34°, is found by adding 2,28;23° and 5,58;11° (= $\lambda + \lambda_S'$). The 'intermediate' epoch used here is the time of the appropriate mean conjunction, and so one can proceed with the use of the tables for finding the corrections. The mean lunar anomaly, 40°, serves as the argument in Table 64 (first row), where the entries are:

$\lambda_L'([0], 40) = 5,57;2°$
$v_L([0], 40) = 0;30,21°/h$
$\omega'([0], 40) = 5,57;2°.$

The true longitude of the Moon, 2,25;25°, is found by adding 2,28;23° and 5,57;2° (= $\bar{\lambda} + \lambda_L'$). The true argument of lunar latitude, 2,17;0°, results from adding 2,19;58° and 5,57;2° (= $\bar{\omega} + \omega'$). There is a difference of 1;9° between the true longitudes of the two luminaries. Dividing this value by $v_L - v_S$ we find 2;28h (text: 2;30h), which is the time between mean and true conjunction. Thus, true conjunction occurs at 11d 6;7h (= 11d, 3;37h + 2;30h) of August 1447. It should be pointed out that the sequence of mean conjunctions, labeled [1] to [12], are based on the mean conjunction labeled [0], which can be the first mean conjunction of the year, but it can also be any other conjunction, in which case we still have a sequence of 13 consecutive conjunctions. The text indicates that this time does not take into account the equation of time. For different approaches to the problem of finding the time between mean and true syzygy by medieval astronomers, see Chabás and Goldstein 1997.

Table 65. Multiplication table

Rc, ff. 80v–82r: *Caput tabula*
Ed. 1495, ff. M5v–O3r: *Tabula proportionum*
Ed. 1526, ff. 307v–321r: *Tabula proportionum*
(This table is not included in Na.)

There are 60 rows corresponding to all integer numbers from 1 to 60, and 59 columns corresponding to all integer numbers from 1 to 59. Moreover, there are additional columns at intervals of 0;10, from 29 to 38. Each entry is the product of the number at the head of the row and the number at the head of the column.

Table 66. Calendaric table for collected years

Na, f. 125v: *Tabula feriarum, aurei numeri et indictionum*
Nu, f. 95r: *Feriarum, aurei numeri et indicionum*
Rc, f. 79r: *Tabula feriarum, aurei numeri et indictionum*
Ed. 1495, f. O3v: *Tabula feriarum, aurei numeri et inditionum*
Ed. 1526, f. 321v: *Tabula feriarum, aurei numeri et inditionum*

The purpose of this table is to determine the weekday, the golden number, and the indiction of a given year (from 2000y to 100y, at intervals of 100y, and from 100y to 0y, at intervals of 10y): see Table 66. The entries in the column headed *ferie* serve as argument for Table 67. The golden number is the number assigned to each year within a cycle of 19 years and such that

golden number ≡ number of the civil year + 1 (mod. 19).

The indiction is the number assigned to each year within a cycle of 15 years such that

indiction − 1 ≡ number of the civil year + 2 (mod. 15).

Table 66: Calendaric table for collected years (excerpt)

Years	28 Ferie	19 Au. nu. Adde 1	15 Inditio Adde 3
2000	12	5	5
1900	24	0	10
...			
20	20	1	5
10	10	10	10
0	0	0	0

Table 67. Calendaric table for expanded years

Na, f. 125v: *Tabula feriarum, aurei numeri et indictionum*
Nu, f. 95r: *Feriarum, aurei numeri et indicionum*
Rc, f. 79r: *Tabula feriarum, aurei numeri et indictionum*
Ed. 1495, f. O4r–v: *Tabula feriarum*
Ed. 1526, f. 322r–v: *Tabula feriarum*

This table gives the Sunday letter of each year in a period of 28 years and the weekday of the first day of each year: see Table 67. The Sunday letter gives the day-number in January for the first Sunday of the year. Thus, in a year where the first day of January is Saturday, the Sunday letter is b (= 2), for the second day of the year is a Sunday. In his copy of Bianchini's tables, Regiomontanus only has columns for the number of years and the Sunday letter.

Table 67: Calendaric table for expanded years (excerpt)

Years		January	February	...	November	December
1	b	Saturday	Tuesday	...	Tuesday	Thursday
2	a	Sunday	Wednesday	...	Wednesday	Friday
3	g	Monday	Thursday	...	Thursday	Saturday
...						
27	c	Wednesday	Saturday	...	Saturday	Monday
28	c	Thursday	Sunday	...	Monday	Wednesday

Table 68. Movable feasts

Na, f. 126r: *Tabula feriarum, aurei numeri et indictionum*
Nu, f. 95r: *Festorum mobilium*
Rc, f. 78v: *Tabula festorum mobilium*
Ed. 1495, f. O5r–v: *Tabula festorum mobilium*
Ed. 1526, f. 323r–v: *Tabula festorum mobilium*

This table gives the date of five movable feasts of the Christian calendar: *Septuagesima* (Septuagesima, the 9th Sunday before Easter), *Quadragesima* (Lent, 40 days before Easter), *Pasca* (Easter), *Ascensio* (Ascension Day, 40 days after Easter), and *Penthecostes* (Whitsunday, 50 days after Easter).

3. OTHER TABLES IN THE MANUSCRIPTS AND ED. 1526 THAT ARE NOT INCLUDED IN ED. 1495

Table 69. Solar equation

Na, f. 15v: *Tabula equationis solis*
Va, f. 187r: *Tabula equacionis solis*

Although the solar equation is already embedded in Table 11, we are given another table with explicit entries for the solar equation. This table, placed between Tables 7 and 8, is also found in other manuscripts containing the Tables of Bianchini, and in the same location.[18] This table also appears in the *editio princeps* of the Alfonsine Tables.[19]

Planetary latitudes: Tables 70 (Saturn), 71 (Jupiter), 72 (Mars), 73 (Venus), and 74 (Mercury)

Na, ff. 118r–119v (superior planets); ff. 120r–121v (inferior planets)
Rc, ff. 76v–77r (superior planets); ff. 77v–78r (inferior planets)
Va, ff. 45r–47v (superior planets); ff. 48r–52v (inferior planets)

---

[18] Rosińska, *Scientific Writings*, 481.
[19] Ratdolt, *Tabule astronomice Alfontij*, e2v–e3v.

Ed. 1526, ff. 324r–326v (Saturn: *Tabula latitudinis* [symbol for Saturn]);
ff. 327r–329v (Jupiter); ff. 330r–332v (Mars); ff. 333r–338v (Venus);
and ff. 339r–344v (Mercury)

Although the planetary latitudes are already embedded in Tables 27, 34,
41, 48, and 56, we are given another set of tables with explicit entries
for the latitude of the planets.[20] This set is found in several manuscripts
containing Bianchini's tables where the superior planets are grouped in
one table and the inferior planets in another. In ed. 1526 there is one
table for each planet.

The tables for Saturn, Jupiter, and Mars have four columns: see Table
70. The first column lists the argument, from 1° to 360° at intervals of
1°. Column 2 is for the minutes of proportion in minutes and seconds,
column 3 displays the northern latitude of the planet in degrees and
minutes, and column 4 gives its southern latitude, also in degrees and
minutes.

Table 70: Latitude of Saturn (excerpt)

[1] Arg. (°)	[2] Min. prop. (min)	[3] Northern (°)	[4] Southern (°)
1	37;36	2; 2	2; 1
2	36;48	2; 2	2; 1
...			
40	0; 0	2;11	2; 7
...			
130	60; 0	2;49	2;51
...			
180	38;24	3; 4	3; 5
...			
220	0; 0	2;54	2;54
...			
310	60; 0	2;13	2;10
...			
360	38;24	2; 2	2; 1

---

[20] Goldstein and Chabás, "Tables for Planetary Latitudes," 457–66.

The columns for northern and southern latitudes are functions of the true anomaly, α, whereas the column for the minutes of proportion is a function of the true center, κ. The columns for the minutes of proportion are the same for the three planets, but for shifts, and they are based on the column in *Almagest* XIII.5 for the minutes of proportion that applies to the latitude for all planets.[21] This particular column in *Almagest* XIII.5 is a function of the 'argument of latitude' counted from the northern limit on the deferent (i.e., 90° from the nodes where the deferent crosses the ecliptic); no algebraic signs are associated with its coefficients. In the instructions to compute the planetary latitudes for the superior planets (*Almagest* XIII.6), Ptolemy indicates that the northern limits on the deferent differ in each case from the apogees of the superior planets by +50° (Saturn), −20° (Jupiter), and 0° (Mars). For example, according to Ptolemy, the ascending node for Saturn is at longitude 90°, i.e., the northern limit (N) is at longitude 180°, and the apogee (A) of Saturn's deferent is 230° (rounded from 233°). Bianchini's table takes these two elements into account: the entries in the column for the minutes of proportion of Saturn are shifted −50° with respect to the corresponding entries in the *Almagest*; those for Jupiter, +20°, and those in the column for Mars have no shift. The inclusion of these shifts makes the tables more 'user-friendly' than Ptolemy's: there is one less step for the user, for he can enter the column for interpolation with the true center in all cases. In other words, there is no longer any need to compute an argument of latitude. Apparently, this was not Bianchini's original idea, for the same shifts can be found in the tables for planetary latitudes associated with Prosdocimo de' Beldomandi of Padua, composed in 1424 (Bologna, University Library, MS 2248, ff. 22v–25v); however, the entries in Prosdocimo's columns other than those for the minutes of proportion differ from those of Bianchini.

The values for the latitude of Saturn range from +2;2° to +3;4° and from −2;1° to −3;5° (see Table 70). Those of Jupiter range from +1;6° to +2;5° and from −1;4° to −2;8°, and those of Mars from +0;5° to +4;21° and from −0;2° to −7;7°.

---

[21] Toomer, *Ptolemy's Almagest*, 632–34.

Chapter 34 includes short comments on planetary latitudes and a worked example for Jupiter. Bianchini considers the true center of the planet to be κ = 126° and its true anomaly, α = 321°. The entries found in Table 71 are 16;24 for the minutes of proportion, a function of true center, and 1;9° for its southern latitude, a function of true anomaly. The resulting latitude is their product: 0;18,51,36° ≈ 0;19°.

The tables for Venus and Mercury have six columns: see Tables 73 and 74. The first column lists the argument from 1° to 360° at intervals of 1°. Column 2 is for the inclination of the planet, here called *declinatio*, in degrees and minutes; column 3 displays the slant of the planet, here called *reflexio*, in degrees and minutes; column 4 gives the minutes of proportion for the inclination, in minutes and seconds; column 5 gives the minutes of proportion for the slant, in minutes and seconds; and column 6 displays the deviation, in degrees and minutes, that is, the third component of latitude.

Table 73: Latitude of Venus (excerpt)

[1] Arg. (°)	[2] Inclination (°)	[3] Slant (°)	[4] M. prop. incl. (min)	[5] M. prop. slant (min)	[6] Deviation (°)
1	1; 3	0; 1	1; 4	59;56	9;59
2	1; 3	0; 2	2; 8	59;52	9;59
...					
45	0;49	1; 1	42;12	42;12	7; 3
...					
90	0; 0	1;57	60; 0	0; 0	0; 0
...					
135	1;45	2;30	42;12	42;12	7; 2
...					
180	7;22	0; 0	0; 0	60; 0	10; 0
...					
225	1;41	2;30	42;12	42;12	7; 3
...					
270	0; 0	1;57	60; 0	0; 0	0; 0
...					
315	0;49	1; 1	42;12	42;12	7; 3
...					
360	1; 3	0; 0	0; 0	60; 0	10; 0

Table 74: Latitude of Mercury (excerpt)

[1] Arg. (°)	[2] Inclination (°)	[3] Slant (°)	[4] M. prop. incl. (min)	[5] M. prop. slant (min)	[6] Deviation (°)
1	1;46	0; 1	1; 4	53;56	44;57
2	1;46	0; 3	2; 8	53;53	44;54
...					
45	1;19	1;22	42;12	37;59	31;39
...					
90	1; 0	2;20	60; 0	0; 0	0; 0
...					
135	2;17	2;17	42;12	46;26	31;39
...					
180	4; 4	0; 0	0; 0	66; 0	45; 0
...					
225	2;17	2;17	42;12	46;26	31;39
...					
270	0; 0	2;20	60; 0	0; 0	0; 0
...					
315	1;19	1;22	42;12	37;59	31;39
...					
360	1;46	0; 0	0; 0	54; 0	45; 0

The columns for the inclination and the slant are functions of the true anomaly, α, whereas those for the minutes of proportion and the deviation are functions of the true center, κ. The columns for the minutes of proportion for the inclination are shifted +90° with respect to the corresponding entries in *Almagest* XIII.5. As for the columns for the minutes of proportion for the slant, that for Venus is the same as that in the *Almagest*, but that for Mercury has no analog in Ptolemy's treatise. Its entries decrease from 54;0′ (at 0°) to 0;0′ (at 90°), increase to 66;0′ (at 180°), decrease to 0;0′ (at 270°), and increase back to 54;0′ (at 360°). There is no indication of the algebraic signs, and they may all be taken to be positive. The extremal values for this correction are indeed $^1/_{10}$ more and $^1/_{10}$ less than 60 minutes, which corresponds to the instructions given by Ptolemy in *Almagest* XIII.6 (but there is no corresponding table in the *Almagest*). It is readily seen that the entries in column 5 for the minutes of proportion of the slant of Mercury can be computed from the minutes of proportion, p(κ), in *Almagest* XIII.5 as follows:

$$c_5(\kappa) = {}^9/_{10} \cdot p(\kappa), \text{ for } 0° \leq \kappa \leq 90°, 270° \leq \kappa \leq 360°$$

and

$$c_5(\kappa) = {}^{11}/_{10} \cdot p(\kappa), \text{ for } 90° \leq \kappa \leq 270°.$$

As for the inclination and slant of the two inferior planets, the entries follow very closely the corresponding values in *Almagest* XIII.5. Now, as regards the deviation for Venus and Mercury, the entries in columns 6 show extremal values of +0;10,0° and –0;45,0°, respectively.[22] It is readily seem that both columns 6 can be computed from $p(\kappa)$:

$$c_6(\kappa) = +0;10 \cdot p(\kappa), \text{ for Venus,}$$

and

$$c_6(\kappa) = -0;45 \cdot p(\kappa), \text{ for Mercury.}$$

Chapter 34 also includes worked examples for Venus and Mercury. In the case of the inferior planets, Bianchini used the terms *declinatio*, *reflexio*, and *deviatio* for the three components of latitude, but we note that in ed. 1495 the first term is *declaratio*, a mistake that was corrected in ed. 1526. We present Bianchini's worked examples without following them word-for-word.

In the example for Venus, the true anomaly $\alpha = 258°$ and the true center $\kappa = 121°$. For this value of α, Table 73 gives $c_2(\alpha) = 0;20°$ (inclination) and $c_3(\alpha) = 2;9°$ (slant). For this κ, Table 73 gives $c_4(\kappa) = 51;24$ (minutes of proportion for the inclination), $c_5(\kappa) = 30;52$ (minutes of proportion for the slant), and $c_6(\kappa) = 0;5,9° = \beta_3$ (deviation). The three components of latitude are $\beta_1 = c_2(\alpha) \cdot c_4(\kappa) = -0;17,8°$, $\beta_2 = c_3(\alpha) \cdot c_5(\kappa) = 1;6,22°$, and $\beta_3 = 0;5,9°$. The latitude of Venus is $\beta = \beta_1 + \beta_2 + \beta_3 = 0;54,23°$, which agrees with the edition of 1495 (although ed. 1526 gives the result as 0;54,22°).

In the example for Mercury, $\alpha = 224°$ and $\kappa = 189°$. Table 74 gives $c_2(\alpha) = 2;20°$ (inclination), $c_3(\alpha) = 2;15°$ (slant), $c_4(\kappa) = 9;24$ (minutes of proportion for the inclination), $c_5(\kappa) = 65;1$ (minutes of proportion for the slant), and $c_6(\kappa) = -0;44,21°$ (deviation). The three components of latitude are $\beta_1 = c_2(\alpha) \cdot c_4(\kappa) = -0;22°$, $\beta_2 = c_3(\alpha) \cdot c_5(\kappa) = -2;26°$, and $\beta_3 = -0;44°$. The latitude of Mercury is $\beta = \beta_1 + \beta_2 + \beta_3 = -3;32°$, and this is the value computed by Bianchini.

---

[22] For an explanation of these parameters, see Neugebauer, *Ancient Mathematical Astronomy*, 222–24.

It should be noted, however, that Bianchini, as well as many other western astronomers, did not follow precisely the instructions given by Ptolemy in the *Almagest*:

$$\beta_3 = p(\kappa) \cdot p(\kappa) \cdot d,$$

where d, the maximum of the third component of latitude for the inner planets (deviation), is 0;10° for Venus and −0;45° for Mercury. Had he followed them, he would have found for Venus β_3 = 0;5,9° · 0;30,52 = 0;2,39°, and the final result would have been β = 0;50,53°. Analogously, for Mercury he would have computed β_3 = −0;44° · 0;59,12 = −0;44°, and in this case his final result would have been the same.

In sum, all the entries in Bianchini's latitudes tables ultimately derive from *Almagest* XIII.5, but Bianchini presented them differently. In particular, he introduced one column for Venus (col. 6) and two columns for Mercury (cols. 5 and 6) which are not found in *Almagest* XIII.5.

Table 75. Lunar equation

Na, f. 122r: *Tabula equationis lune*
Rc, f. 75v: *Tabula equationis lune*
Va, f. 259v: *Tabula equacionis lune*
Ed. 1526, ff. 345r–347v: *Tabula equationis* [symbol for the Moon]

Although the lunar equation is already embedded in Table 17, we are given another table with explicit entries for the lunar equation. This is the same table that appears in the *editio princeps* of the Alfonsine Tables.[23]

Planetary equations: Tables 76 (Saturn), 77 (Jupiter), 78 (Mars), 79 (Venus), and 80 (Mercury)

Na, f. 122v (Saturn: *Tabula equationis Saturni*); f. 123r (Jupiter); f. 123v (Mars); f. 124r (Venus); and f. 124v (Mercury)
Rc, f. 74v: (Saturn: *Tabula equationis Saturni*); ff. 74v–75r (Jupiter); f. 75r (Mars); ff. 75v–76r (Venus); and f. 76r (Mercury)
Va, f. 260r: (Saturn: *Tabula equacionis Saturni*); f. 260v (Jupiter); f. 261r (Mars); f. 261v (Venus); and f. 262r (Mercury)

---

[23] Ratdolt, *Tabule astronomice Alfontij*, e4r–e6v.

Ed. 1526, ff. 348r–350v (Saturn: *Tabula equationis* [symbol for Saturn]);
ff. 351r–353v (Jupiter); ff. 354r–356v (Mars); ff. 357r–359v (Venus);
and ff. 360r–362v (Mercury)

Although the planetary equations are already embedded in Tables 27,
34, 41, 48, and 56, we are given another set of tables with explicit entries
for the planetary equations. These are the same tables that appear in
the *editio princeps* of the Alfonsine Tables.[24]

Table 81. Stations

Na, f. 125r: *Tabula stationum 5 planetarum*
Rc, f. 78v: *Tabula stationum 5 planetarum*
Va, f. 262v: *Tabula stationum quinque planetarum*
Ed. 1526, ff. 363r–365v: *Tabula stationum quinque planetarum*

This table has 6 columns. The first lists the arguments at intervals of 1°,
from 1° to 3,0°. There is one column for the first stations of each of the
first stations of the five planets. According to the headings in ed. 1526,
the units are degrees and minutes, but this is a mistake for physical
signs and degrees. This table is an expanded version of those for the
same purpose in *Almagest* XII.8 and the zij of al-Battānī[25] where the
entries are given with a higher precision (to minutes), but at intervals
of 6° of the argument. We note that no such table appears in the *editio
princeps* of the Alfonsine Tables.

Table 82. Equation of time

Na, f. 108r–v: *Tabula horarum meridiei et equationis dierum ad meridianum ferarrie et bononie*
Rc, f. 48r–v: *Tabula horarum meridiei ad meridianum ferarrie et bononie
et equationis dierum*
Va, ff. 258v–259r: *Tabula horarum meridiei et equationis dierum ad
meridianum ferarrie et bononie*
Ed. 1526, f. 366r–v: *Tabula equationis dierum*

[24] Ratdolt, *Tabule astronomice Alfontij*, e7r–g5v.
[25] Nallino, *Al-Battānī*, 2:138–39.

In the manuscript tradition[26] the equation of time is presented together with the length of daylight (Table 83), and this is indeed the case in Na, Rc, and Va. We shall treat the two items separately.

The entries in this table are given in minutes and seconds of time, whereas in most tables of this kind the units are degrees and minutes: see Table 82. To transform the entries of a table for the equation of time from units of arc to units of time, one has to multiply the value by 4, for 360 time-degrees = 24 hours or 1 time-degree = 0;4 hours. This explains why all entries in Table 82 are multiples of 4. The extremal values in this table are:

> max = 22;12 min (Tau 28° – Gem 5°)
> min = 12;16 min (Leo 1° – 9°)
> Max = 31;36 min (Sco 8° – 9°)
> Min = 0;0 min (Aqu 18° – 19°).

This table is a variant of the table for the equation of time associated with the Toledan Tables,[27] with a maximum of 7;54°, and yielding a maximum value, expressed in time, of 31;36 min (= 7;54 · 4). In the manuscripts containing the Toledan Tables, the equation of time is usually combined with right ascensions in a single table, and this is also the case in the zij of al-Battānī, where the same table is found.[28] Toomer and Pedersen located copies of this table with only the equation of time, expressed in minutes of an hour.[29] In the *editio princeps* of the Alfonsine Tables, the equation of time (in degrees) was combined with right ascensions in a single table.[30]

Table 82: Equation of time (excerpt)

Arg. (°)	Ari (min)	Tau (min)	Gem (min)	Cnc (min)	Leo (min)	Vir (min)
1	9; 0	18;44	22;12	17; 4	12;16	15;12
2	9;20	18;56	22;12	16;56	12;16	15;24
...						
29	18;16	22;12	17;36	12;24	14;48	23;56
30	18;32	22;12	17;20	12;20	15; 0	24;56*

* instead of 24;16.

---

[26] See Rosińska, *Scientific Writings*, 485.
[27] Pedersen, *Toledan Tables*, 968–85.
[28] Nallino, *Al-Battānī*, 2:61–64.
[29] Toomer, "Toledan Tables," 35; Pedersen, *Toledan Tables*, 977–80.
[30] Ratdolt, *Tabule astronomice Alfontij*, k1r–k2r.

Table 82 (*cont.*)

Arg. (°)	Lib (min)	Sco (min)	Sgr (min)	Cap (min)	Aqu (min)	Psc (min)
1	24;36	31;16	27;56	14;44	2;28	0;48
2	24;56	31;20	27;40	14;12	2;12	1; 0
...						
29	31; 8	28;20	15;48	2;56	0;32	8;20
30	31;12	28; 8	15;16	2;44	0;40	8;40

It is interesting to note that another tradition for the equation of time developed in the 14th and 15th centuries, and it is preserved in tabular form in several manuscripts. This table is usually ascribed to John of Lignères and exhibits a maximum of 7;57°, but actually it was already described by Peter of St. Omer, a predecessor who also worked in Paris, in a text dated 1293–1294.[31] In Figure 4, we represented the curves corresponding to both traditions: al-Battānī–Toledan Tables–*editio princeps* of the Alfonsine Tables (here called 'Ratdolt') and Peter of St. Omer–John of Lignères (here called 'J. of L.').

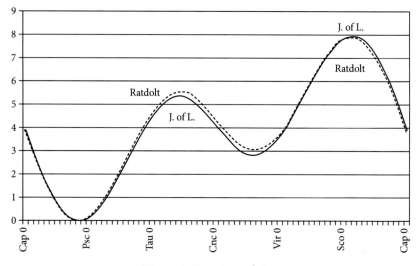

Figure 4: Equation of time.

---

[31] Pedersen, *Toledan Tables*, 984.

Table 83. Length of daylight for Ferrara and Bologna

Na, f. 108r–v: *Tabula horarum meridiei et equationis dierum ad meridianum ferarrie et bononie*

Rc, f. 48r: *Tabula horarum meridiei ad meridianum ferarrie et bononie et equationis dierum*

Va, ff. 258v–259r: *Tabula horarum meridiei et equationis dierum ad meridianum ferarrie et bononie*

Ed. 1526, f. 367r–v: *Tabula horarum meridiei ad meridianum Ferrarie ac Bononie*

The argument in this table is the true solar longitude, λ, given at intervals of 1°: see Table 83. The entries represent the difference between 24h and the half-length of daylight, $M/2$, given in hours, minutes, and seconds. We note that all entries in this table are multiples of 4, which may be a sign that they were adapted from a similar table, in degrees. The maximum and minimum entries (19;45,12h and 16;14,48h, respectively) are mutually consistent, for 19;45,12h = 18h + 1;45,12h and 16;14,48h = 18h – 1;45,12h. This implies that the maximum length of daylight is 15;30,24h (= 2 · (24 – 16;14,48)) and the minimum length of daylight is 8;29,36h (= 2 · (24 – 19;15,12)). The entries in the table can be computed by means of the modern formula

$$24h - M/2 = 24h - (90 + \arcsin(\tan \Delta(\lambda) \cdot \tan \varphi))/15,$$

where φ is the geographical latitude of the observer, $\Delta(\lambda) = \arcsin(\sin \lambda \cdot \sin \varepsilon)$, and ε is the obliquity of the ecliptic. Bianchini does not say how the entries in this table were computed, but the best results with this modern formula are obtained when $\varepsilon = 23;51°$ and the latitude of Ferrara and Bologna is $\varphi = 45°$, as indicated in the Introduction, and not $\varphi = 44;45°$, as stated in the heading of Table 86. It is surprising that here Bianchini uses Ptolemy's value for the obliquity rather than one of the lower values available at Bianchini's time: 23;35°, the value in al-Battānī's zij;[32] 23;33,30°, the value in the Toledan Tables;[33] and 23;30°, used by Regiomontanus to compute the half-length of daylight in his *Kalendarium*, published in 1483.[34] In a passage in his

---

[32] Nallino, *Al-Battānī*, 2:58.
[33] Toomer, "Toledan Tables," 30–31; Pedersen, *Toledan Tables*, 961.
[34] Chabás and Goldstein, *Abraham Zacut*, 155.

Introduction concerning precession/trepidation, Bianchini cites the parameters 23;51° in the name of Ptolemy and 23;33,30° in the name of al-Ma'mūn. However, in a treatise ascribed to Thābit Ibn Qurra, *On the motion of the eighth sphere*, extant only in a Latin version, it is said that "observers at the time of [the caliph] al-Ma'mūn [d. 833] found 23;33° [for the obliquity]".[35] The parameter 23;33,30° has been ascribed to Azarquiel.[36]

In sum, it is likely that Bianchini borrowed this specific table from an earlier set of tables computed for his latitude.

Table 83: Complement to the half-length of daylight (excerpt)

Arg.(°)	Ari(h)	Tau(h)	Gem(h)	Cnc(h)	Leo(h)	Vir(h)
1	17;58,24	17;10,48	16;31,36	16;14,48	16;33,12	17;13,36
2	17;56,48	17; 9,12	16;30,48	16;14,48	16;34,24	17;15,12
...						
29	17;13,36	16;33,12	16;14,48	16;31,36	17;10,48	17;58,24
30	17;12,24	16;32,24	16;14,48	16;32,24	17;12,24	18; 0, 0

Arg.(°)	Lib(h)	Sco(h)	Sgr(h)	Cap(h)	Aqu(h)	Psc(h)
1	18; 1,36	18;49,12	19;28,24	19;45,12	19;26,24	18;46,24
2	18; 3,12	18;50,48	19;29,12	19;45,12	19;25,36	18;44,48
...						
29	18;46,24	19;27,24	19;45,12	19;28,24	18;49,12	18; 1,36
30	18;41,36	19;27,36	19;45,12	19;27,56	18;47,36	18; 0, 0

Rc, f. 48r, has a table, similar to the combination of Tables 82 and 83, entitled *Tabula horarum meridiei ad meridianum Romanum et equationis dierum*, but there are no entries in the column for the complement of the length of daylight.

---

[35] Neugebauer, "Thābit ben Qurra," 293.
[36] Pedersen, *Toledan Tables*, 962.

Table 84. Domification for the 5th climate

Na, ff. 109r–111v: *Tabula equationis domorum in climate quinto*
Nu, ff. 101v–104r: *Aequacio domorum in climate quinto ubi scilicet dies prolixior anni est horarum 15*
Rc, ff. 64v–65r, *Tabula equationis 12 domorum ad meridianus quinti climatis cuius elevatio poli est 40;55° et longitudo longioris diei 15;0h*
Ed. 1526, ff. 368r–373v: *Tabula equationis domorum in 5 climate* (although the pages are in the correct order, those following 370 are labeled '365', '372', and '367')

Tables 84, 85, and 86 are for the 'equation of the houses', i.e., for 'domification'. Each table is for a different climate where the term, climate, refers to a geographical latitude. Traditionally, there were seven climates.[37] There are 12 sub-tables in each table, one for each of the 12 zodiacal signs. Here, we only display excerpts of two sub-tables, for the Sun in Aries and the Sun in Cancer: see Table 84. These headings are misleading since the longitude of the Sun is not the argument for any of the columns (but see Table 85A).

Each sub-table has 9 columns. The first column displays, at intervals of 1°, the degree in the zodiacal sign in the heading, and it is the argument for the next column only, labeled *merid(ies)*. The entries in the column labeled *merid(ies)* are right ascensions, counted from Cap 0°, as a function of the solar longitude, $\alpha'(\lambda)$, called 'normed right ascension'. The entries are given in hours and minutes, whereas the usual tables for *sphaera recta* have entries in degrees and minutes. The entries in this column were taken from a table for right ascensions, $\alpha(\lambda)$, by shifting the tabulated values according to the rule $\alpha'(\lambda) = \alpha(\lambda) + 90°$, and converting the result into units of time. So, for example, the entry for 30° is 7;52h, and its equivalent, 117;53°, occurs in the table for right ascensions in the Toledan Tables.[38] Although $15 \cdot 7;52 = 118$ (rather than 117;53), Bianchini probably rounded it to 118°, the entry he found in the Toledan Tables, and derived his entry for 30° from it. Alternatively, Bianchini computed $118/15 = 7;51,32$, and rounded it to 7;52. The entries in the column headed *meridies* are the same for all climates.

[37] See, e.g., Toomer, *Ptolemy's Almagest*, 19.
[38] Pedersen, *Toledan Tables*, 973.

In the next column, headed *horolo(gium)*, the entries are the oblique ascensions, $\rho(\lambda)$, expressed in hours and minutes, where the argument is the entry in the column labeled '1', i.e., the *horoscopus* (not the degree of longitude in the zodiacal sign that appears in the heading, as it was for the preceding column). The entries in this column are equivalent to those for the 5th climate in the Toledan Tables.[39] So, for example, for the 5th climate the entry in column 1 for 30° in the sub-table, 'The Sun in Aries', is Lib 30° = 210°, and the corresponding entry in the column labeled *horologium* is 14;33h. In the Toledan Tables, $\rho(210°)$ is 218;9°= 14;33h, as in our text.[40] We note that there are no tables of right or oblique ascensions in either edition of Bianchini's tables, and that the tables for right and oblique ascensions in the *editio princeps* of the Alfonsine Tables are the same as those in the Toledan Tables.[41] Again, we emphasize that the column labeled *meridies* is properly associated with the zodiacal sign in the heading, but the column labeled *horologium* is properly associated with the columns that come after it, not with the zodiacal sign in the heading.

The purpose of the two columns, labeled *meridies* and *horologium*, is to allow the user of these tables to convert the time on a given day (represented by the true solar longitude on that day) to the *horoscopus*, the point on the ecliptic that is rising at that moment which is the cusp of the first astrological house. The true Sun crosses the meridian at true noon, and so we need to compute the right ascension of the true Sun in order to find the point on the equator that crosses the meridian at true noon of that day. We then add the time since noon to get the point on the equator that crosses the meridian at the time in question. In Figure 5, HM is the arc on the ecliptic that corresponds to arc EC (= 90°) on the equator. To get the longitude of point H, we would need to add 90° to the position of point C, but this is already included in the function $\alpha'(\lambda)$. So, with the sum of the normed right ascension of the true Sun and the time since noon, we look in the table of oblique ascensions, seeking the longitude that corresponds to this amount, and it is the longitude of the point H. Note that EC is always 90°, but that HM in general is not 90°. The length of arc EC follows from the fact that ES = 90° and EZ = 90°, which implies that E is the pole of great circle SCZ,

---

[39] Pedersen, *Toledan Tables*, 1057–59.

[40] Pedersen, *Toledan Tables*, 1059.

[41] Ratdolt, *Tabule astronomice Alfontij*, k1r–l4v; Pedersen, *Toledan Tables*, 968–75, 1036–72.

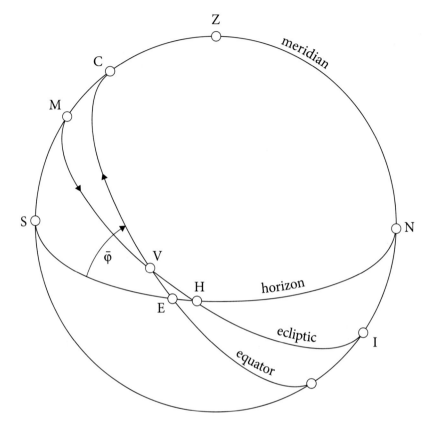

Figure 5: Domification.

and the distance from the pole to any point on its great circle is always 90°. The longitude of M, the cusp of the 10th house, is found from the inverse right ascension function, that is, α(M) = arc VC (counted in the direction of increasing longitude). But with normed right ascension, the rule is α'(M) = ρ(H).[42]

For example, let the longitude of the true Sun be 60°, and the time after true noon be 4h (= 60°). We seek the cusps of houses 1 and 10 at that time. First we need to find the normed right ascension of the true Sun: α'(60°) = 147;47°, and then we add the right ascension corresponding to the time of day to it: the result is 207;47°. Next we seek the inverse oblique ascension of 207;47° for the fifth climate and, in the

---

[42] Neugebauer, *Ancient Mathematical Astronomy*, 42.

Toledan Tables,[43] it is 201;54° (with interpolation): this is the cusp of house 1, or the *horoscopus*, H, for the time in question. Note that the cusp is the point on the ecliptic marking the beginning of each astrological house. To get the longitude of M, we need the oblique ascension of H for the fifth climate, $\rho(201;54°) = 207;47°$, and its inverse normed right ascension, 115;46°, is the longitude of M, the cusp of house 10. The cusp of house 4 is diametrically opposite the cusp of house 10; hence cusp 4 is 295;46°. In the table in ed. 1526 for the fifth climate, where the ascendant (cusp 1) is Lib 21° (= 201°), cusp 4 is 294;42° and, interpolating, where the ascendant is Lib 21;54° (= 201;54°), cusp 4 is 295;47°, in close agreement with the recomputation.

The column headed 1 gives the longitude of the ascendant, λ_1, in degrees and minutes, and has the label *Horoscopus* at the bottom (written out in full in ed. 1526, but abbreviated in Tables 84–86 as *Hor.*). The entries in the rest of the columns, headed 2 to 6, represent the longitudes λ_i (where i = 2,…, 6), also in degrees and minutes, of the cusps as a function of the longitude λ_1 of the ascendant. The cusps of houses 7,…, 12 can be derived from the entries in the table by setting $\lambda_j = \lambda_{j-6} + 180°$ (where j = 7,…, 12).

The Libra zero test, as it has been called by North,[44] is an easy way to determine the obliquity of the ecliptic embedded in these tables. In all three of them, when λ_1 = Lib 0° = 180°, the longitudes of the cusps are $\lambda_2 = 212;12°$, $\lambda_3 = 242;7°$, $\lambda_4 = 270°$, $\lambda_5 = 297;53°$, $\lambda_6 = 327;48°$; thus, $\lambda_7 = 0°$, $\lambda_8 = 32;12°$, $\lambda_9 = 62;7°$, $\lambda_{10} = 90°$, $\lambda_{11} = 117;53°$, and $\lambda_{12} = 147;48°$. These values indicate that the obliquity used in the computation of the tables is 23;33°.[45]

Tables of domification were a standard component of zijes.[46]

---

[43] Pedersen, *Toledan Tables*, 1059.
[44] North, *Horoscopes and History*, 14–15.
[45] Pedersen, *Toledan Tables*, 1076.
[46] See, e.g., Pedersen, *Toledan Tables*, 1075–1121; Zacut, *Almanach Perpetuum*, 18r–20v; Suter, *Al-Khwārizmī*, 194–205; Neugebauer, *Al-Khwārizmī*, 128–129; cf. North, *Horoscopes and History*.

Table 84: Equation of the houses for the 5th climate (excerpt)

	merid.	horolo.	1 Lib	2 Sco	3 Sgr	4 Cap	5 Cap	6 Aqu
(°)	(h)	(h)	(°)	(°)	(°)	(°)	(°)	(°)
				Sun in Aries				
1	6; 4	12; 5	1; 0	3;16	3;12	1; 9	28;59	28;53
							Aqu	
2	6; 7	12;10	2; 0	4;20	4;17	2;19	0; 6	29;58
...								
				Sgr	Cap	Aqu	Psc	Ari
29	7;48	14;27	29; 0	2;15	3;12	4;30	1;27	0; 9
30	7;52	14;33	30; 0	3;15	4;16	5;44	2;40	1;18
			Hor.					

	merid.	horolo.	1 Cap	2 Aqu	3 Psc	4 Ari	5 Tau	6 Gem
(°)	(h)	(h)	(°)	(°)	(°)	(°)	(°)	(°)
				Sun in Cancer				
1	12; 4	19;34	1; 0	6;12	14;53	25;26	18;33	10;13
2	12; 9	19;38	2; 0	7;15	16; 3	26;32	19;36	11;14
...								
				Psc	Ari	Tau	Gem	Cnc
29	14; 5	21;22	29; 0	5;46	15; 8	22;44	15;22	6;56
30	14; 9	21;25	30; 0	6;48	15; 8	23;35	16; 4	7;59
			Hor.					

Table 85. Domification for the 6th climate

Na, ff. 112r–114v: *Tabula equationis domorum ad (…) Ferrarie et Bononie climatis 6 cuius latitudo est 45;24°*

Nu, ff. 104v–107r: *Aequaciones domorum in climate sexto ubi scilicet dies prolixior anni est horarum 15 cum media*

Rc, ff. 65v–66r, *Tabula equationis 12 domorum ad meridianus sexti climatis cuius elevatio poli est 45;24° et longitudo longioris diei 15;30h*

Ed. 1526, ff. 374r–379v: *Tabula equationis domorum ad meridianum Ferrarie et Bononie in 6 climatis cuius latitudo est 44;45°*

The geographical latitudes explicitly given in the titles differ, and for this reason we have reproduced excerpts of both tables: see Tables 85 and 85A. A point of interest is that, for astronomers at the time, 45;24° was the latitude assigned to Padua, whereas those for Ferrara and Bologna were 45;0° and 44;30°, respectively. These are the values found in the Alfonsine Tables compiled in 1424 by Prosdocimo de' Beldomandi, and preserved in Bologna, University Library, MS 2248.[47] However, in his *Tabulae primi mobilis*, as well as in his first letter to Regiomontanus dated 1463, Bianchini gives an accurate value of the latitude of Ferrara, 44;45,4°.[48] An example of the confusion associated at that time with geographical coordinates is that in the *editio princeps* of the Alfonsine Tables printed in Venice, which is quite close to these three places, the latitudes assigned to Padua, Ferrara, and Bologna are 45°, 44;0°, and 44;30°, respectively. Thus, it is likely that the latitude given in the Naples manuscript is that of Padua whereas the latitude in ed. 1526 corresponds to Ferrara.

Table 85: Equation of the houses for the 6th climate (MSS Na, Nu, and Rc)

			Sun in Aries					
			1	2	3	4	5	6
	merid.	*horolo.*	Lib	Sco	Sgr	Cap	Cap	Aqu
(°)	(h)	(h)	(°)	(°)	(°)	(°)	(°)	(°)
1	6; 4	12; 5	1; 0	3;18	3;15	1;13	29; 2	28;54
							Aqu	
2	6; 7	12;11	2; 0	4;14	4;23	2;26	0;11	0; 1
...								
				Sgr	Cap	Aqu	Psc	Ari
29	7;48	14;34	29; 0	2;52	4;14	6; 2	2;38	0;51
30	7;52	14;39	30; 0	3;53	5;20	7;20	3;54	2; 2
			Hor.					

			Sun in Cancer					
			1	2	3	4	5	6
	merid.	*horolo.*	Cap	Aqu	Psc	Ari	Tau	Gem
(°)	(h)	(h)	(°)	(°)	(°)	(°)	(°)	(°)
1	12; 4	19;49	1; 0	7;25	17;25	29;26	21; 3	11;23
						Tau		
2	12; 9	19;53	2; 0	8;28	18;45	0;32	22; 6	12;23
...								
				Psc	Ari		Gem	Cnc
29	14; 5	21;33	29; 0	6;46	17;17	25;43	17; 3	7;50
30	14; 9	21;37	30; 0	7;51	18;17	26;31	17;55	8;46
			Hor.					

---

[47] Chabás, "Prosdocimo de' Beldomandi."
[48] Zinner, *Regiomontanus*, 38, 61.

In the table published in ed. 1526 the entries differ slightly from those in the manuscripts as a result of the different value used for the geographical latitude. Moreover, the layout is not quite the same, for an extra column was added at the far left, for the day in a year, beginning with March 12. (In 1448 the vernal equinox occurred shortly before noon March 11, and it is thus consistent to have March 12 correspond to Ari 1° in Table 85A.) The extra column in Table 85A makes sense of the heading (as is not the case in the other tables for domification), for the date in this column corresponds to the solar longitude when the Sun is in the zodiacal sign indicated in the heading and in the degree found in the corresponding row.

Table 85A: Equation of the houses for the 6th climate (ed. 1526)

				Sun in Aries					
				1	2	3	4	5	6
		merid.	horolo.	Lib	Sco	Sgr	Cap	Cap	Aqu
Day	(°)	(h)	(h)	(°)	(°)	(°)	(°)	(°)	(°)
March 12	1	6; 4	12; 5	1; 0	3;17	3;14	1;12	29; 1 Aqu	28;54
March 13	2	6; 7	12;10	2; 0	4;22	4;21	2;25	0; 9	30; 0
...					Sgr	Cap	Aqu	Psc	Ari
April 10	29	7;48	14;33	29; 0	2;46	4; 3	5;46	2;43	0;45
April 11	30	7;52	14;39	30; 0 Hor.	3;47	5; 9	7; 3	4; 2	1;55

				Sun in Cancer					
				1	2	3	4	5	6
		merid.	horolo.	Cap	Aqu	Psc	Ari	Tau	Gem
Day	(°)	(h)	(h)	(°)	(°)	(°)	(°)	(°)	(°)
June 14	1	12; 3	19;46	1; 0	7;12	17; 6	28;43	20;36	11;11
June 15	2	12; 6	19;51	2; 0	8;16	18;16	29;49	21;39	12;12
...					Psc	Ari	Tau	Gem	Cnc
July 13	29	14; 4	21;31	29; 0	6;37	16;56	25;11	16;45	7;42
July 14	30	14; 9	21;35	30; 0 Hor.	7;38	17;54	25;58	17;34	8;38

Chapter 27 is devoted to the casting of a horoscope for Ferrara and offers a worked example. We are told that the Sun is in Sgr 17;20° at 18;24h, counted from true noon. In Table 85, sub-table for the Sun in Sagittarius, we find 23;3h for 17° and 23;8h for 18°, in the column headed *merid.* By interpolation, the normed right ascension, expressed in hours and minutes, corresponding to the given position is 23;4,40h. We are then told to add it to the given time and to subtract 24h from the sum. The result is 17;28,40h. This value is the entry in the column headed *horolo.*, i.e., the oblique ascension, expressed in hours and minutes. The closest values in this column are 17;24h and 17;29h, corresponding to the ascendants Sgr 1° (= 241°) and Sgr 2° (= 242°), respectively, found in the sub-table for the Sun in Gemini. In the Toledan Tables, $\rho(241°)$ = 260;57° = 17;24h, and $\rho(242)$ = 262;15° = 17;29h, in agreement with the text.[49] We note that Table 85A (ed. 1526) has different values, 17;22h and 17;27h, indicating, first, that Bianchini computed with Table 85 and not Table 85A and, second, that he was faithful to the Alfonsine tradition which, in this case, depended on the Toledan Tables. The longitudes of the cusps found in Table 85 in the row for Gem 1° are λ_1 = Sgr 1°, λ_2 = Cap 5;45°, λ_3 = Aqu 11;9°, λ_4 = Psc 20;13°, λ_5 = Ari 14;51°, and λ_6 = Tau 8;40°. If Bianchini had used Table 85A, he would have found values differing from those given explicitly in the text. The difference in time between the given value and that found in the table (0;4,40h = 17;28,40h – 17;24h) requires interpolation in all houses. The final results for the cusps of the first 6 houses are λ_1 = Sgr 1;56°, λ_2 = Cap 6;45°, λ_3 = Aqu 12;17°, λ_4 = Psc 21;32°, λ_5 = Ari 16;3°, and λ_6 = Tau 9;45°. Bianchini's canons do not indicate how to compute the cusps of the astrological houses, but North (1986) describes various ways to do so. We have recomputed these values, where ε = 23;33° and φ = 45;24°, using a modified version of the computer program developed by North, which he kindly provided to us. The results are λ_1 = Sgr 1;56,0°, λ_2 = Cap 6;42,1°, λ_3 = Aqu 12;18,18°, λ_4 = Psc 21;35,4°, λ_5 = Ari 16;4,0°, and λ_6 = Tau 9;43,47°, in reasonably good agreement with the text.

Table 86. Domification for the 7th climate

Na, ff. 115r–117v: *Tabula equationis domorum climatis 7*
Nu, ff. 98v–101r: *Aequaciones domorum in climate septimo ubi scilicet dies anni prolixior est horarum 16*

---

[49] Pedersen, *Toledan Tables*, 1064.

Rc, ff. 66v–67r, *Tabula equationis 12 domorum ad meridianus septimi climatis cuius elevatio poli est 48;40° et longitudo longioris diei* [blank]
Ed. 1526, ff. 380r–385v: *Tabula domorum 7 climatis in quo est Austria, Cracconia et Polonia cuius latitudo est 50;11°*

The layout of this table is the same as that of Table 84: see Table 86. The topographical names in the heading are not found in the manuscripts, and we assume that they were added in ed. 1526 for some unstated reason. 'Cracconia' is probably a misprint for Cracovia (= Cracow), spelled 'Crascovia' in Table 87, below, with coordinates 44;20° (longitude) and 50;0° (latitude). These are the same coordinates found in the list of geographical coordinates in the *editio princeps* of the Parisian Alfonsine Tables, where this city is correctly called 'Cracovia'.

Table 86: Equation of the houses for the 7th climate

			Sun in Aries					
			1	2	3	4	5	6
	merid.	*horolo.*	Lib	Sco	Sgr	Cap	Cap	Aqu
(°)	(°)	(h)	(h)	(°)	(°)	(°)	(°)	(°)
1	6; 4	12; 5	1; 0	3;18	3;16	1;15	29; 4	28;56
							Aqu	Psc
2	6; 7	12;11	2; 0	4;24	4;25	2;30	0;15	0; 3
...								
				Sgr	Cap	Aqu	Psc	Ari
29	7;48	14;39	29; 0	3;18	5; 9	7;32	3;40	1;20
30	7;52	14;45	30; 0	4;20	6;16	8;53	4;58	2;32
			Hor.					

			Sun in Cancer					
			1	2	3	4	5	6
	merid.	*horolo.*	Cap	Aqu	Psc	Tau	Tau	Gem
(°)	(h)	(h)	(°)	(°)	(°)	(°)	(°)	(°)
1	12; 4	20; 5	1; 0	8;39	20;15	3;16	23;28	12;33
2	12; 9	20; 9	2; 0	9;43	21;24	4;20	24;29	13;33
...								
				Psc	Ari		Gem	Cnc
29	14; 5	21;46	29;0	7;59	19;36	29;46	19; 2	8;52
30	14; 9	21;49	30;0	9; 0	20;32	29;31	19;50	9;45
			Hor.					

Rc, ff. 67v–68r, has a table for domification for Florence, for a latitude
of 43;44° and a longest daylight of 15;19h.

4. OTHER TABLES IN ED. 1526 OR IN MS NU THAT ARE NOT INCLUDED IN ED. 1495 OR IN MS NA

Table 87. Geographical coordinates

Ed. 1526, f. 1r–v: No title

In this table we are given the geographical coordinates (longitude and
latitude, both in degrees and minutes) of 80 places arranged, for the
most part, by increasing longitude. The westernmost place is Córdoba
(9;0° and 37;0°) and the easternmost is *Civitas venti* (83;0° and 31;0°)
which we have not identified. By contrast, in the *editio princeps* of the
Parisian Alfonsine Tables the westernmost place is *Fortunae insulae*,
i.e., The Fortunate Islands, or the Canary Islands as they are now
called (3;0° and 11;0°), and the easternmost place is *Civitas reg. altini*
(150;30° and 18;30°) which probably corresponds to *civitas regni acim*
in the Toledan Tables (i.e., a city in China), based on an Arabic version
of the coordinates of cities in Ptolemy's *Geography*.[50] The Latin *acim*
comes from the Arabic term for China, *al-Ṣīn* or *aṣ-Ṣīn*, where the
'n' has become an 'm' and the letter 'ṣ' is represented by 'c', as occurs
frequently in medieval Latin transliterations of Arabic. It is also likely
that the 'c' and the 'm' in *acim* were read as 't' and 'ni' by the person
responsible for the archetype of the list in the Alfonsine Tables (variants
of this kind are not uncommon in medieval Latin), thus yielding *altini*
instead of *acim*. The latitude of the place according to the Toledan Tables
is 18;30° (as in the Alfonsine Tables), but the longitude according to
the Toledan Tables is 177;0° in contrast to the Alfonsine Tables where
it is 150;30°, a discrepancy that we cannot explain. All except four of
the places in the list in ed. 1526 are also found in the *editio princeps* of
the Parisian Alfonsine Tables with the same coordinates (but for a few
cases). The four localities added in ed. 1526 are Valencia (14;15° and
39;0°), Lucca (33;0° and 42;31°), Vicenza (33;10° and 44;30°), and *Mons
Gaurus*, given the same coordinates as Salerno (38;20° and 41;0°). The

---

[50] Pedersen, *Toledan Tables*, 1512, no. 14: note that the Arabic tradition differs from
the readings in the Greek manuscripts.

coordinates of Toledo are 11;0° and 41;0°. Ferrara (33;30° and 44;30°), Rome (35;20° and 42;0°), and Naples (37;0° and 40;20°) have special marks before their names.

Below the first part of the table (f. 1r) a later hand has added that the difference between Ferrara and Barcelona is 1;5h.

The use of this table is explained in Chapter 1 where an example is given for a place whose longitude is 40;30° (no such place is listed in the table). The example corroborates that the longitude of Ferrara is 32°, as stated in the introduction, and not 33;30°, as in this list.

Table 88. Solar equation and solar velocity at mean conjunction

Ed. 1526, ff. 301v–304r: *Tabula* [symbol for the Sun] *in* [symbol for conjunction]

In this table the argument is the mean solar anomaly, κ̄, given from 1° to 360° at intervals of 1°: see Table 88. Here the *locus* is the standard solar equation used in Alfonsine astronomy, but presented in a way to make it always positive. The extremal values for the solar velocity are: 0;2,23°/h for κ̄ = 1°,…, 35° and 325°,…, 360°, and 0;2,34°/h for κ̄ = 139°,…, 190°. For a similar table, with the argument given in days of a year, see Table 11. The entries in this table are identical to those appearing in the first row of each sub-table of Table 63.

Table 88: Solar equation and solar velocity at mean conjunction (excerpt)

Anom. (°)	Locus (°)	ad horas (°/h)
1	5,59;58	0;2,23
2	5,59;56	0;2,23
...		
93	5,57;50	0;2,28
...		
180	0; 0, 0	0;2,34
...		
267	2;10, 0	0;2,28
...		
359	0; 0, 2	0;2,23
360	0; 0, 0	0;2,23

Table 89. Lunar equation and lunar velocity at mean conjunction

Ed. 1526, ff. 304v–307r: *Tabula* [symbol for the Moon] *in* [symbol for conjunction]

In this table the argument is the mean lunar anomaly, $\bar{\alpha}$, given from 1° to 360° at intervals of 1°: see Table 89. Here the *locus* is the standard lunar equation used in Alfonsine astronomy but, by adding 360° to the negative values, it is presented in a way to make all values positive. The extremal values for the lunar velocity are: 0;29,37°/h for $\bar{\alpha}$ = 1°, 2° and 360°, and 0;36,52°/h for $\bar{\alpha}$ = 176°,..., 180°. The entries in this table are almost identical to those appearing in the first row of each sub-table of Table 64. A similar velocity table, with the same extremal values, is found twice in Oxford, Bodleian Library, MS Can. Misc. 499 (ff. 41v–42r, 154v–155r) where it is ascribed to John of Lignères, and in a printed edition of 1536.[51] The entries for the lunar velocity in the table ascribed to John of Lignères were computed by means of the formula:

$$v(\alpha) = 0;32,56 + 0;41,49 \cdot \Delta,$$

where 0;32,56°/h is the lunar hourly mean motion in longitude, 0;41,49°/h is the corrected hourly lunar mean motion in anomaly at syzygy, and $\Delta = c_6(\alpha + 1) - c_6(\alpha)$.

This table and that ascribed to John of Lignères differ on the argument to be used: the argument is the mean lunar anomaly whereas for John of Lignères it is the true lunar anomaly. Moreover, this table is not symmetric about 180°.

The table ascribed to John of Lignères (and similar ones) appear in manuscripts of the 14th and 15th centuries, but the underlying formula is not given by anyone in the 14th century, as far as we know. However, the equivalent of the previous formula is given in words in a text printed at the end of the 15th century.[52] In the 14th and 15th centuries this formula was only used for computing the lunar velocity at syzygy and was not applied at other elongations.

The author of this table tried to adapt the procedure described by this formula so that the argument would be $\bar{\alpha}$ rather than α. Although he (Bianchini?) does not explain his procedure, we found good numerical

---

[51] Schöner, *Tabulae Astronomicae*.

[52] Regiomontanus, *Epythoma*, VI,4. For further details, see Goldstein, "Lunar Velocity." In addition to the copies of this table cited there, we can now add those of John of Gmunden (d. 1442): see Porres, *Jean de Gmunden*, 329–33.

agreement between a formula on which we based our reconstruction and the entries in his table. As can be seen in Table 89, the differences between the entries do not differ greatly from those in the table ascribed to John of Lignères, but the differences are significant and systematic. The formula used by in this case, as we reconstruct it, is:

$$v(\bar{\alpha}) = 0;32,56 + 0;41,49 \cdot \Delta$$

where

$$\Delta = c_6(\bar{\alpha} + 1 + c_6(\bar{\alpha} + 1)) - c_6(\bar{\alpha} + c_6(\bar{\alpha})),$$

and the amount to be added to the argument, $c_6(\bar{\alpha})$, is always taken to be positive. For example, for $\bar{\alpha} = 58°$, $c_6(59°) = -4;2,17°$ and of $c_6(58) = -3;59,26°$. Hence, with the supplement to the argument taken as positive, $c_6(59° + 4;2,17°) = c_6(63;2,17°) = -4;13,9°$, $c_6(58° + 3;59,26°) = c_6(61;59,26°) = -4;10,25°$, and $\Delta = -4;13,9° + 4;10,25° = -0;2,44°$. It follows that

$$v(\bar{\alpha}) = 0;32,56 - 0;41,49 \cdot 0;2,44 = 0;31,2°/h.$$

In this table $v(58°) = 0;31,6°/h$, whereas in the table ascribed to John of Lignères $v(58°) = 0;30,57°/h$. To take another example, where $\bar{\alpha}$ is greater than 180°: let $\bar{\alpha} = 302°$. Then $c6(303°) = 3;56,30°$, and $c_6(303° + 3;56,30°) = c_6(306;56,30°) = 3;44,21°$; and $c6(302°) = 3;59,26°$, $c_6(302° + 3;59,26°) = c_6(305;59,26°) = 3;47,22°$; hence $\Delta = 3;44,21° - 3;47,22° = -0;3,1°$. It follows that

$$v(302°) = 0;32,56 - 0;41,49 \cdot 0;3,1 = 0;30,50°/h.$$

In this table $v(302°) = 0;30,45°/h$ which differs from the value in his table for $v(58°)$, whereas in the table ascribed to John of Lignères $v(302°) = 0;30,57°/h$ which is the same value that is given for $v(58°)$. A less precise computation, without interpolation in c_6, yields even closer agreement with the entries in this table.

In our view, the formula, which we assume was used by its author, is unjustified. In fact, the difference between $\bar{\alpha}$ and α is not $c_6(\bar{\alpha})$ but $c_3(2\eta)$. And a direct computation of the lunar progress in an hour subsequent to mean syzygy yields a value for lunar velocity that is not the same as the entry in the table. For this computation we need to take the increments in 1 hour in mean double elongation, $2\eta = 1;1°$, in mean longitude, $\bar{v} = 0;32;56°$, and in mean anomaly, $\bar{\alpha} = 0;32,40°$. We then compute the progress in longitude since mean syzygy, that is, the difference between the true longitude of the Moon at syzygy and the true longitude of the Moon 1 hour later. If the mean anomaly at mean syzygy is 58°, the true lunar longitude at syzygy (relative to its mean position at syzygy) is $c_6(\bar{\alpha}) = c_6(58) = -3;59,26°$. After 1 hour, $c_3(2\eta) = c_3(1;1°) = 0;9°$, $c_4(2\eta) = c_4(1;1°) = 0$, and $\bar{\alpha} = 58;32,40°$. Hence $\alpha = \bar{\alpha} + c_3 = 58;32,40° + 0;9° = 58;41,40°$, $c_6(58;41,40°) = -4;1,22°$. Since the lunar correction according the standard Alfonsine Tables is $c(\bar{\alpha}, 2\eta) = c_5(\alpha) \cdot c_4(2\eta) + c_6(\alpha)$ and, in this case, $c_4(2\eta) = 0$, there is no correction to $c_6(\alpha)$. So in 1 hour the Moon has progressed by an amount, $v(58°) = 0;32,56° - 4;1,22° + 3;59,26° = 0;31,0°$, whereas here we have $v(58°) = 0;31,6°$. Our reconstruction does not yield exact agreement, and it is possible that another procedure can be found that produces better agreement. But we think it unlikely that this procedure, whatever it was, can be justified astronomically.

Table 89A: Lunar equation and lunar velocity at mean syzygy (excerpt)

[1]	[2]	[3]	[4]	[5]
		Bianchini	J. of Lignères	
anom.	Locus	ad horas	v(a)	[3] – [4]
(°)	(°)	(°/h)	(°/h)	sec.
1	5,59;55	0;29,37		
...				
30		0;30, 3	0;29,59	4
...				
60		0;31,11	0;31, 3	8
...				
90		0;32,51	0;32,40	11
...				
95	5,55; 4	0;33,10		
...				
120		0;34,47	0;34,36	11
...				
150		0;36,21	0;36,18	3
...				
180	0; 0	0;36,52	0;36,53	–1
...				
210		0;36, 3	0;36,18	–15
...				
240		0;34,20	0;34,36	–16
...				
265	4;56	0;32,44		
...				
270		0;32,24	0;32,40	–16
...				
300		0;30,50	0;31, 3	–13
...				
330		0;29,53	0;29,59	–6
...				
360	0; 0	0;29,37	0;29,37	0

Table 90. Excess of revolution for the Sun in expanded years

Nu, f. 96r: *Tabula revolucionum annorum*
Rc, f. 79r: *Tabula de revolutionibus annorum*
Ed. 1526, f. 386r: *Tabella annuarum conversionum post bisextum anni*

Tables 90–93 are used to determine the time, t, at which the Sun has the same position as it had at a given epoch, t_0, before the time sought, that is, after very nearly an integer number of tropical years has elapsed. Chapter 40 gives two worked examples for the use of these tables: for times close to the epoch, that is, when the solar apogee does not change significantly, and for times well after the epoch. Different tables apply in these two cases.

The argument of this table is the number of years from 1y to 40y at intervals of 1y: see Table 90. In addition to the column for the argument, there are 4 columns, one for each year in a 4 year-cycle, beginning with a leap year. One enters the table in the row which has the number of years since the epoch and then looks in the column for the character of the year of the epoch, i.e., if it is a leap year, one looks in the column so labeled, and if it is not a leap year one looks at the entry in the appropriate column. The entries, in hours, minutes, and seconds, are the excesses over 365 days after a number of years have elapsed. Thus, in the first row we are told to add (A) 5;49,16h to the time of the epoch when one year has elapsed or to subtract (M) 18;10,44h (= 24h – 5;49,16h) from it when the year of the epoch is the third year after a leap year. We note that 5;49,15,36h derives from the standard Alfonsine tropical year-length of 365d 5;49,15,36h = 365;14,33,6d. In the second row we are told to add (A) 11;38,32h (= 2 · 5;49,16h) to the time of the epoch when two years have elapsed, in the case of an epoch that was in a leap year or 1 year after a leap year, or to subtract 12;21,18h (= 24h – 11;38,32h) from it when the year of the epoch is 2 or 3 years after a leap year. In row 4 (and multiples of 4) the increment in time is the same in all four columns (4 · 5;49,16h = 23;17,4h = 24h – 0;42,56h), and each entry in row 40 is then 10 times the entry in row 4.

Table 90: Excess of revolution for the Sun in expanded years (excerpt)

Years	Leap year (h)	1 (h)	2 (h)	3 (h)
1	5;49,16 A	5;49,16 A	5;49,16 A	18;10,44 M
2	11;38,32 A	11;38,32 A	12;21,18 M	12;21,28 M
3	17;27,48 A	6;32,12 M	6;32,12 M	6;32,12 M
4	0;42,56 M	0;42,56 M	0;42,56 M	0;42,56 M
...				
40	7; 9,20 M	7; 9,20 M	7; 9,20 M	7; 9,20 M

Table 91. Excess of revolution for the Sun in a year

Nu, f. 96r: *Aequacio revolucionum*
Rc, f. 79r: *Equatio revolutionum*
Ed. 1526, f. 386v: *Tabula equatonis* [*revolutionibus*] [symbol for the Sun]

As is the case for most other tables above, this table is unprecedented in the astronomical literature, as far as we know: see Table 91. The argument is the true solar longitude, given in degrees at intervals of 1°. The entries, in seconds of an hour, represent the correction, whether negative (*mi.*) or positive (*adde*), to be applied to the time given in the preceding table when the Sun returns to the same true longitude after approximately 1 year.

In the first worked example presented in Chapter 40 we are asked to consider as epoch Nov. 8, 1383 at 18;38,4h. We are told that the true position of the Sun is Sco 24;59,39° = 234;59,39°. The problem is to find the time in year 1435 when the Sun was in the same position as it was in the year of the epoch. These two dates are relatively close, for they are only separated by 52 years. In Table 88 we find the excesses for 40y (7;9,20h) and 12y (2;8,48h). Both entries are found in the column headed 3, corresponding to the third year after a leap year, which is the case for 1383 and 1435. Both entries are assigned the letter M, for *minue* (subtract). Subtracting the sum of these two entries (9;18,8h) from the epoch, we obtain 9;19,56h. This time has to be corrected for the specific solar position (Sco 24;59,39° ≈ Sco 25°) whose entry in Table 91 is 0;0,26h, with the notation *adde* (add). We are then told to multiply this correction by the number of years elapsed, and to add the result, 0;22,32h, to the time found previously, 9;19,56h. The new result, 9;42,28h, corresponds to the time on Nov. 8, 1435 when the Sun was in the same position as in the year of the epoch.

Table 91: Equation of the excess of revolution for the Sun in a year

	Ari mi.	Tau mi.	Gem mi.	Cnc mi.	Leo mi.	Vir mi.	Lib mi.	Sco adde	Sgr adde	Cap adde	Aqu adde	Psc adde
0	0	17	28	33	30	19	3	14	29	34	29	17
1	1	17	28	33	29	18	2	15	29	34	29	16
2	1	18	29	33	29	18	2	15	29	34	29	16
3	2	18	29	32	29	17	1	16	29	33	28	15
4	2	18	29	32	28	17	1	16	29	33	28	14
5	3	19	29	32	28	16	adde	17	30	33	28	14
6	3	19	29	32	27	16	0	17	30	33	27	13
7	4	19	29	32	27	15	1	18	30	33	26	13
8	4	20	29	32	27	15	2	18	30	33	26	12
9	5	20	30	32	26	14	2	19	30	33	26	12
10	6	21	30	32	26	14	3	19	31	32	25	11
11	6	21	30	32	26	13	3	20	31	32	25	11
12	7	21	30	31	25	12	4	20	31	32	24	10
13	7	22	30	31	25	12	5	21	31	32	24	9
14	8	22	31	31	25	11	5	21	31	32	23	9
15	8	23	31	31	24	11	6	22	31	32	23	8
16	9	23	31	31	24	10	7	22	32	32	23	8
17	9	23	31	31	23	10	7	23	32	31	22	7
18	10	24	31	31	23	9	8	23	32	31	22	7
19	11	24	31	31	23	9	8	24	32	31	21	6
20	11	24	31	31	22	8	9	24	32	31	21	6
21	12	25	31	31	22	8	9	24	32	31	20	5
22	12	25	32	30	22	7	10	25	33	31	20	4
23	13	26	32	30	21	7	10	25	33	30	20	4
24	13	26	32	30	21	6	11	26	33	30	19	3
25	14	26	32	30	21	6	11	26	33	30	19	3
26	15	27	32	30	20	5	12	27	33	30	18	2
27	15	27	32	30	29	5	13	27	33	30	18	2
28	16	28	32	30	19	4	13	28	34	30	18	1
29	16*	28	33	30	19	4	14	28	34	30	17	1

* Ed. 1526 has 0. The typesetter omitted one entry in this column, causing an upwards shift of all subsequent entries and leaving nothing for the last row; hence he entered a '0'. The other entries in this column have silently been corrected to agree with the entries in the manuscripts.

Bianchini's canons give no hint of the way this table was constructed. The following method to recompute the entries is entirely based on Bianchini's tables but, although yielding good results, we cannot claim that this was the method used by the author. First, we should keep in mind that, as can be seen in Table 8, the yearly progress of the solar apogee ranges from 0;0,55° (around epoch) to 0;0,20° (around year 2000), and it is about 0;0,34° in Bianchini's time. Whatever the yearly increment, $\Delta\lambda_A$, it implies a variation of the solar equation in a year, Δe, which in turn depends on the solar position within the year. To determine Δe corresponding to $\Delta\lambda_A$, let us consider two instants approximately 1 year apart, when the true solar longitude returns to the same value, λ. Between these two instants the true anomalies differ in $\Delta\lambda_A$. If we take the mean anomalies to differ also in $\Delta\lambda_A$, then Δe can be derived by interpolation from a table for the solar equation for the value of the mean anomaly, $\bar{\kappa}$, corresponding to a true anomaly, $\kappa = \lambda - \lambda_A$. As can be seen in Table 8, in Bianchini's time λ_A was about 90;40°. The value obtained for Δe then has to be converted into hours by dividing it by the hourly solar velocity taken from Table 11. Thus, for example, for $\lambda = 60°$ and for years 1444 and 1445, when the solar apogee was at 1,30;39,8° and 1,30;39,42°, as indicated in Table 8, $\Delta\lambda_A$ = 0;0,34h. The corresponding true solar anomalies were 329;20,52° and 329;20,18°. For these values of anomaly approximately 1 year apart, the increment in solar equation, Δe, is 0;0,1,3°, as derived from the table for the solar equation. Now, the solar velocity, v, for a solar anomaly of 329° is 0;2,23°/h, as indicated in Table 11. Thus, $\Delta e/v$ is = 0;0,27h (text: 0;0,28h). It is entirely possible that the author only computed a few selected entries and filled in the rest by interpolation.

Table 92. Excess of revolution for the Sun in collected years

Nu, f. 96v: *Tabula revolucionum secunda*
Rc, f. 79v: *Tabula prima solis*
Ed. 1526, ff. 387r: *Tabula prima* [symbol for the Sun] *in revolutionibus annorum*

This table gives the number of days, hours, minutes, and seconds that exceeds a number of years of 365 days, for collected years from 40y to 2000y at intervals of 40y. The first entry, for 40y, is 0d 7;9,21h, and differs by 1 second from that in Table 90. Curiously, the heading of this table in Regiomontanus's copy reads 'in annis expansis' instead of 'in annis collectis'.

Table 93. Excess of revolution for the Sun due to solar anomaly

Nu, f. 96v: *Tabula revolucionum secunda*
Rc, f. 79v: *Tabula 2 in revolutionibus*

Ed. 1526, ff. 387v–388r: *Tabula secunda* [symbol for the Sun] *in revolutionibus annorum*

This table gives the number of days, hours, minutes, and seconds as a function of the solar anomaly, from 1° to 180°, and from 180° to 359°, at intervals of 1°, where each argument and its complement in 360° are given in the same row: see Table 93A. The maximum, 2d 4;46,17h, occurs at anomaly 92°–94°. We have recomputed the entries in this table by finding the quotient of the solar correction in the Parisian Alfonsine Table divided by the mean motion of the Sun, 0;59,8,19,37°/d, the value in these tables. The differences between text and computation as shown in Table 93A are less than 0;1h in all but one case, and differences of this magnitude have no significance here, although we are aware that the column for the differences, T(ext) – C(omputation), exhibits some sort of a sinusoidal component. We note that the computer of this table did not consider a variable motion for the Sun. The entry is the time it takes the Sun to travel the distance in longitude between its mean position and its true position as a function of the solar anomaly.

In the second worked example presented in Chapter 40 we are asked to consider as epoch Aug. 25, year 141 at 4h. We are told that the mean longitude of the Sun at that time was 2,33;5,1°, and that the solar apogee, the solar anomaly, and the true longitude of the Sun were 1,13;27,40°, 1,19;37,21°, and 2,30;58,35°, respectively. The problem here is to find the time of year 1450 when the Sun was in the same position as at the epoch. These two dates are separated by 1309 years. In Table 92 we find the excess for 1280y (9d 12;59,37h) and in Table 90 that for 29y (0;48,44h). The entry for 1280y has the letter M, for *minue* (subtract). The entry for 29y is to be found in the column headed 1, corresponding to the first year after a leap year, as is the case for year 141, and has the letter A, for *adde* (add). Subtracting the first entry from the epoch and adding the second entry, we obtain Aug. 15 at 15;49,7h. This time has now to be corrected because of the change in solar anomaly. In order to do this, we are asked to determine in Table 93 the excesses corresponding to two values of the solar anomaly. The first is the anomaly at epoch (1,19;37,21°) for which the entry in Table 93 is 2d 3;22,32h, after interpolation (2d 3;22,25h according to the worked example). The second is the anomaly that results from subtracting the mean longitude of the Sun at epoch (2,33;5,1° = 1,13;27,40° + 1,19;37,21°) from the value of the solar apogee for 1450 (1,30;42,33°, as deduced from Table 8), and it

Table 93A: Excess of revolution for the Sun due to solar anomaly

Anomaly (°)	Excess (d, h)		Comp. (d, h)		Diff. (T – C) (h)
10	0	8;42,37	0	8;42,42	–0; 0, 5
...					
20	0	17;20,17	0	17;20, 8	0; 0, 9
...					
30	1	1;31,40	1	1;31,35	0; 0, 5
...					
40	1	8;48,18	1	8;47,26	0; 0,52
...					
50	1	15;22,19	1	15;21,54	0; 0,25
...					
60	1	20;56,50	1	20;56,18	0; 0,32
...					
70	2	0;52,46	2	0;52,30	0; 0,16
...					
80	2	3;26,47*	2	3;26,18	0; 0,29
...					
90	2	4;45, 4	2	4;44,13	0; 0,51
...					
92–94	2	4;46,17	2	4;45,26	0; 0,51
...					
100	2	4;15,50	2	4;15, 0	0; 0,50
...					
110	2	2;21,30	2	2;21,22	0; 0, 8
...					
120	1	22;39,47	1	22;38,59	0; 0,48
...					
130	1	17;22,43	1	17;22,56	–0; 0,13
...					
140	1	10;56,25	1	10;55,16	0; 1, 9
...					
150	1	3;14,12	1	3;14,15	–0; 0, 3
...					
160	0	18;37,20	0	18;37,14	0; 0, 6
...					
170	0	9;36,56	0	9;37, 5	–0; 0, 9
...					
180	0	0; 0, 0	0	0; 0, 0	0; 0, 0

* Ed. 1526 has [0d] 23;26,47h. This is an isolated printer's error: the entries before and after it are 2d 3;15,25h and 2d 3;37,45h, respectively.

is 1,2;22,28°. The corresponding entry in Table 93 is 1d 22;0,22h, after interpolation (1d 22;0,6h according to the worked example). Following the text, the difference is 0d 5;22,19h (= 2d 3;22,25h – 1d 22;0,6h) and it has to be subtracted from the time computed above (Aug. 15 at 15;49,7h). The result is Aug. 15 at 10;26,48h, which corresponds to the time in 1450 when the Sun was in the same position it had in the year of the epoch.

Table 94. Excess of revolution for the Moon and the ascending node in collected years

Nu, f. 97r: *Lune et capitis in revolucionibus annorum collectorum*
Rc, f. 79v: *Tabula lune in revolutionibus annorum*
Ed. 1526, ff. 388v–389r: *Tabula* [symbol for the Moon] *et* [symbol for the ascending node] *in revolutionibus annorum collectorum*

In this table there are six columns: see Table 94. The first is for the argument: collected years from 40 to 2000 at intervals of 40y. The entries in the other columns represent the time (in days, hours, and minutes), called [*dies*] *superabundantes*; the double elongation (in degrees and minutes), called *centrum*; the mean longitude of the Moon (in physical signs, degrees, and minutes), called *motus*; the mean argument of lunar latitude (in physical signs, degrees, and minutes); and the longitude of the lunar ascending node (in physical signs, degrees, and minutes). In Regiomontanus's copy the column for the argument of lunar latitude was omitted. The time displayed in column 2 is the excess in days over an integer number of anomalistic months; hence, the entries in this column do not exceed 27d 13;18,36h, which is the length of the anomalistic month used in Bianchini's tables (see Table 12, above). Tables 94 and 95 are similar in layout to Tables 12 and 13, but the entries differ, as will be explained below.

Table 94: Excess of revolution for the Moon and the ascending node in
collected years (excerpt)

Years	Time (d, h)		Double elongation (°)	Longitude (°)	Arg. latitude (°)	Lunar node (°)
40	5	18;51	25;20	3, 6;58	4, 0;19	5, 6;21
80	11	13;42	50;41	13;56	2, 0;57	4,12;42
...						
1960	7	22;31	39;50	3,12; 6	5, 0;20	3,11;21
2000	13	17;22	65;10	19; 4	3, 0;39	2,17;42

In Rc, f. 79v, Tables 94 and 95 are merged together.

*Table 95. Excess of revolution for the Moon and the ascending node in
expanded years*

Nu, f. 97r: *Lune et capitis in revolucionibus annorum expansorum*
Ed. 1526, ff. 389v: *Tabula* [symbol for the Moon] *et* [symbol for the
ascending node] *in revolutionibus annorum collectorum*

The layout of this table is the same as that of Table 94, but here the
argument is the number of expanded years, from 1y to 40y: see Table
95. The entries for 40 expanded years agree with those for 40 collected
years in Table 12. As was the case in Table 94, in Regiomontanus's copy
the column for the argument of lunar latitude was omitted.

Table 95: Excess of revolution for the Moon and the ascending node in
expanded years (excerpt)

Years	Time (d, h)		Double elongation (°)	Longitude (°)	Arg. latitude (°)	Lunar node (°)
1	7	0;48	93;41	39;54	58;52	5,40;41*
2	14	1;36	187;21	1,19;49	1,57;45	5,21;19
...						
39	26	7;22	339;50	2,24; 0	2,56;55	5,25;42
40	5	18;51	25;20	3, 6;58	4, 0;19	5, 6;21**

* Ed. 1526 has 5,40;40°.
** Ed. 1526 has 5,6;1°.

The entry for 1y, 7d 0;48h, means that in a year of 365 days 5;49,16h there is an integer number of anomalistic months plus 7d 0;48h and, indeed, 7d 0;48h = 365d 5;49,16h − (13 · 27d 13;18,36h). This entry differs from that for 1y given in Table 13, 6d 18;59h, which was the difference between a year of 365 days and the duration of 13 anomalistic months. The entries for the other four quantities (double elongation, mean lunar longitude, mean argument of lunar latitude, and longitude of the lunar node) have to be understood as the increments in these quantities between the beginning of the year and this precise moment (7d 0;48h before the completion of a year of 365d 5;49,16h).

Table 96. Mean motion of the lunar node in hours

Nu, f. 97r: *Capitis in horis et fracciones*
Rc, f. 80r, no title
Ed. 1526, f. 390r: *Motus* [symbol for the ascending node] *ad horas*

This table has the same layout as Table 7 and it gives the mean motion of the Sun and the lunar node for a number of hours. In this case, the entries represent the difference between 360° and the mean motion of the lunar node for a number of hours (from 1h to 60h at intervals of 1h). Hence, the entry for 1h is 0;0° and that for 60h is 5,59;52°.

Excess of revolution for the planets in collected years: Tables 97 (Saturn), 98 (Jupiter), 99 (Mars), 100 (Venus), and 101 (Mercury)

Nu, f. 97v: *In revolucionibus annorum collectorum*
Ed. 1526, f. 390v (Saturn: *Tabula* [symbol for Saturn] *in revolutionibus annorum collectorum*); f. 391r (Jupiter); f. 391v (Mars); f. 392r (Venus); and f. 392v (Mercury)

In Regiomontanus's copy this information is displayed in a single table in contrast to ed. 1526 where it is presented in different tables, following the same pattern as that for the mean motion of the planets in collected years (Tables 21, 28, 35, 42, and 49). For each of the planets there are three columns: see Table 97. The first is for the argument: collected years from 40 to 2000 at intervals of 40y. The entries in the other columns represent the time (in days, hours, and minutes), with no specific heading, and the increment in longitude (superior planets) or in anomaly (inferior planets), in physical signs, degrees, and minutes. Both these increments are here called *motus* and correspond to the number of collected years. The time displayed in column 2 is the excess over an integer number of periods of anomaly where the length of the year is 365d 5;49,16h.

Table 97: Mean motion of Saturn in collected years (excerpt)

Years	Time (d, h)		*Motus* (°)
40	242	5; 7	2, 1;16
80	106	8; 0	4,15;12
...			
1400	159	10;16	3,22;53*
...			
2000	11	16;49	5,48;30

* Regiomontanus's copy has no row labeled 1400, for the rows corresponding to 1360, 1400, and 1440 were incorrectly labeled 1440, 1480, and 1520.

Comparing this table with Table 21 we note that their corresponding entries for the *motus* are identical, whereas those for the time differ. The explanation is that the time in Table 21 refers to a year of 365d while that in Table 97 refers to a year of 365d 5;49,16h. However, as the *motus* is the increment in anomaly after a number of periods of anomaly in a certain number of years, both tables display the same entries for the same number of collected years.

Excess of revolution for the planets in expanded years: Tables 102 (Saturn), 103 (Jupiter), 104 (Mars), 105 (Venus), and 106 (Mercury)

Nu, f. 97v: *In revolucionibus annorum expansorum*
Ed. 1526, f. 393r (Saturn: *Tabula* [symbol for Saturn] *in revolutionibus annorum expansorum*); f. 393r (Jupiter); f. 393v (Mars); f. 393v (Venus); and f. 394r (Mercury)

The layout of this table is the same as that of Tables 97–101, but here the argument is the number of expanded years, from 1y to 40y: see Table 102. The presentation is the same as that for the mean motion of the planets in expanded years (Tables 22, 29, 36, 43, and 50). In Regiomontanus's copy all the information is gathered in a single table. The entries for 1y are 365d 5;49h (time) and 0;0° (*motus*) for Saturn,

Jupiter, Mars, and Venus; and 17d 14;34h (time) and 5,42;39° (*motus*) for Mercury. As was the case with the tables for collected years, the entries for time differ from those in Tables 22, 29, 36, 43, and 50, whereas the entries for the *motus* are identical in both sets of tables.

Table 102: Mean motion of Saturn in expanded years (excerpt)

Years	Time (d, h)		Motus (°)
1	365	5;49	0, 0; 0
2	352	9;26	0,12;40
3	339	13; 3	0,25;20
...			
40	242	5; 7	2, 1;16*

* Ed. 1526 has 2,1;1°.

In each case, the entry for 2 years results from combining the entry for 1 year and the period of anomaly of the planet concerned. Thus, in the case of Saturn, 352d 9;26h = 2 · 365d 5;49h − 378d 2;12h, where 378d 2;12h is its period of anomaly (see Table 24). The entries for 40 expanded years for each planet agree with those for 40 collected years in Tables 95–99, respectively.

In Rc, f. 80r, Tables 97–101 and 102–106 are merged together.

Table 107. Animodar

Ed. 1526, ff. 394v–395v: *Tabula more infantis in utero. Per D. Jacobuum de Dondis patavinum*

In this table the argument, here called *distantia*, is the argument of lunar latitude and it is given at intervals of 1° from 1° to 180°: see Table 107. The entries are distributed in two columns headed *mora occidentalis* and *mora orientalis*, both given in days, hours, and minutes. They represent the duration of pregnancy with a minimum of 259d 7;50h (in excess of 37 weeks) and a maximum of 288d 0;0h (in excess of 41 weeks). One sees immediately that the entries in the first column have constant differences of 1;50h, whereas in the second column the constant difference is 2;10h (except for the first 12 entries, which are all 273d 0;0h). We note that for a given *distantia* greater than 12° the sum of the entries in these two columns is an integer number of days.

Table 107: Animodar (excerpt)

Distantia (°)	Mora occidentalis (d, h)		Mora orientalis (d, h)	
1	259	7;50	273	0; 0
2	259	9;40	273	0; 0
3	259	11;30	273	0; 0
...				
178	272	20;20	287	19;40
179	272	22;10	287	21;50
180	273	0; 0	288	0; 0

This table is attributed in the heading to Jacopo de Dondi (1298–1359) of Padua, the author of a set of tables of which no copy has yet been properly identified. Despite this attribution, we have found the same table in a zij preserved in Latin, *al-Muqtabis* by Ibn al-Kammād (early 12th century), extant in Madrid, MS 10023, ff. 62v–64r.[53] In both tables the first 12 entries for *mora orientalis* are all 273d 0;0h. This and similar tables, including the one found in the work of Abraham Zacut in the second half of the 15th century, belong to an astrological tradition on nativities that goes back to Ptolemy.[54]

Chapter 50 provides explanations for using this table as well as an example for July 20, 1438.

Table 108. Mean motion of the Moon in days

Ed. 1526, ff. 396r: *Motus* [symbol for the Moon] *in diebus more*

Tables 108 and 109 are associated with the animodar (Table 107), and even the term *mora* (i.e., duration) appears in the headings of the columns in both of them. Table 108 gives the mean lunar longitude and the mean lunar anomaly for a number of days between 258 and 288. It is similar to Table 18 for the daily mean motion of the Moon where we are also given the mean motion of the argument of lunar latitude. We note that, in this table, physical signs are also used for the lunar anomaly, contrary to Bianchini's usage elsewhere.

---

[53] Chabás and Goldstein, "Ibn al-Kammād."
[54] Chabás and Goldstein, *Abraham Zacut*, 150–153.

Table 108: Mean motion of the Moon in days (excerpt)

Days	Longitude (°)	Anomaly (°)
258	5, 9;31	4,10;46
259	5;22;41	4,23;50
...		
287	6, 1;37	4,29;39
288	6;14,48	5,12;43

Table 109. Mean motion of the Moon in hours

Ed. 1526, ff. 396r: *Motus* [symbol for the Moon] *in hora*

This table has the same layout as that of Table 108 and, up to 24 hours, it reproduces the entries for the lunar longitude and the lunar anomaly displayed in Table 19: see Table 109.

Table 109: Mean motion of the Moon in hours (excerpt)

Hours	Longitude (°)	Anomaly (°)
1	0;32,56	0;32,40
2	1; 5,53	1; 5,19
...		
29	12;37,39	12;31,14
30	13;10,35	13; 3,54

Table 110. Mean motion of the Sun in hours

Ed. 1526, ff. 396v–397r: *Tabula motus* [symbol for the Sun] *in qualibet hora per L. Gauricum*

As the headings indicate, Tables 110 and 111 were not computed by Bianchini, but by Luca Gaurico, who was responsible for ed. 1526. The entries in Table 110 represent the distance (in minutes and seconds) traveled by the Sun in a number of hours (from 1h to 24h, at intervals of 1h) at a given velocity (from 0;2,23°/h to 0;2,34°/h, at intervals of 0;0,1°/h).

Table 110: Mean motion of the Sun in hours (excerpt)

Hours	0;2,23 (′)	0;2,24 (′)	...	0;2,33 (′)	0;2,34 (′)
1	2;23	2;24		2;33	2;34
2	4;46	4;48		5; 6	5; 8
...					
23	54;49	55;12		58;30	59; 2
24	57;12	57;36		61;12	61;36

Table 111. Mean motion of the Sun in minutes of an hour

Ed. 1526, ff. 397v: *Tabula motus* [symbol for the Sun] *in m. horarum per Gauricum supputata*

The entries in this table represent the distance (in minutes and seconds) that the Sun travels in a number of minutes of an hour (from 0;5h to 0;55h, at intervals of 0;5h) at a given velocity (from 0;2,23°/h to 0;2,34°/h, at intervals of 0;0,1°/h): see Table 111.

Table 111: Mean motion of the Sun in minutes of an hour (excerpt)

Min	0;2,23	0;2,24	...	0;2,33	0;2,34
5	0;0,12	0;0,12		0;0,13	0;0,13
10	0;0,24	0;0,24		0;0,26	0;0,26
...					
50	0;1,59	0;2, 0		0;2, 8	0;2, 9
55	0;2,11	0;2,12		0;2,21	0;2,21

Table 112. Conversion of degrees into hours

Ed. 1526, f. 1r: *Tabula convertendi g. et m.*

The purpose of this table is to convert time-degrees (given here at 1°-intervals from 1° to 60°) into time (in hours and minutes), where 360° = 24h.

NOTATION

α	true argument of anomaly
ᾱ	mean argument of anomaly
α	right ascension
α′	normed right ascension
β	latitude
Δ	increment
η	elongation
ε	obliquity of the ecliptic
κ	true argument of center
κ̄	mean argument of center
κ	true solar anomaly
κ̄	mean solar anomaly
λ	true longitude
λ̄	mean longitude
M	length of daylight
μ	daily mean motion
p	precession
q	lunar equation
ρ	oblique ascension
φ	terrestrial latitude
v	hourly velocity
v̄	mean hourly velocity
ω	true argument of latitude
ω̄	mean argument of latitude

REFERENCES

Alfonsine Tables. See Ratdolt, *Tabule astronomice Alfontij*; Santritter, *Tabule Astronomice Alfonsi Regis*; and Chabás and Goldstein, *Alfonsine Tables of Toledo*.

Bianchini, Giovanni: *Tabulae astronomiae* (Venice, 1495).

——: *Tabule Joa[nni] Blanchini Bononiensis* (Venice, 1526).

Boffito, Giuseppe: "Le Tavole astronomiche di Giovanni Bianchini da un codice della Collezione Olschki," *La Bibliofilia*, 9 (1907–1908), 378–388 and 446–460.

Chabás, José: *El Lunari de Bernat de Granollachs: Alguns aspectes de la història de l'astronomia a la Catalunya del Quatre-cents* (Barcelona, 1985).

——: "The Astronomical Tables of Jacob ben David Bonjorn," *Archive for History of Exact Sciences*, 42 (1991), 279–314.

——: "Astronomy in Salamanca in the Mid-Fifteenth Century: The *Tabulae Resolutae*," *Journal for the History of Astronomy*, 29 (1998), 167–175.

——: "The Diffusion of the Alfonsine Tables: The case of the *Tabulae Resolutae*," *Perspectives on Science*, 10 (2002), 168–178.

——: "From Toledo to Venice: The Alfonsine Tables of Prosdocimo de' Beldomandi," *Journal for the History of Astronomy*, 38 (2007), 269–281.

Chabás, José, and Bernard R. Goldstein: "Andalusian Astronomy: *al-Zīj al-Muqtabis* of Ibn al-Kammād," *Archive for History of Exact Sciences*, 48 (1994), 1–41.

——: "Computational Astronomy: Five Centuries of Finding True Syzygy," *Journal for the History of Astronomy*, 28 (1997), 93–105.

——: *Astronomy in the Iberian Peninsula: Abraham Zacut and the Transition from Manuscript to Print*. Transactions of the American Philosophical Society, 90.2 (Philadelphia, 2000).

——: *The Alfonsine Tables of Toledo*. Archimedes: New Studies in the History and Philosophy of Science and Technology, 8 (Dordrecht and Boston, 2003).

——: "Early Alfonsine Astronomy in Paris: The Tables of John Vimond (1320)," *Suhayl*, 4 (2004), 207–294.

Curtze, Maximilian: "Der Briefwechsel Regiomontan's mit Giovanni Bianchini, Jacob von Speier und Christian Roder," *Abhandlungen zur Geschichte der mathematischen Wissenschaften*, 12 (1902), 187–336.

Delambre, Jean Baptiste Joseph: *Histoire de l'astronomie du moyen âge* (Paris, 1819).

Dobrzycki, Jerzy: "The *Tabulae Resolutae*," in *De Astronomia Alphonsi Regis*, ed. Mercè Comes, Roser Puig, and Julio Samsó (Barcelona, 1987), 71–77.

Federici Vescovini, Graziella: "Bianchini, Giovanni," *Dizionario biografico degli Italiani*, 10 (1968), 194–196.

Garuti, Paolo: "Giovanni Bianchini: Compositio instrumenti (Cod. Lat. 145 T.6.19) della Biblioteca Estense di Modena," *Rendiconti Classe di Lettere e Scienze Morali e Storiche*, 125 (1992), 95–127.

Gerl, Armin: *Trigonometrisch-astronomisches Rechnen kurz vor Copernicus: Der Briefwechsel Regiomontanus-Bianchini*. Boethius: Texte und Abhandlungen zur Geschichte der exakten Wissenschaften, 21 (Stuttgart, 1989).

Goldstein, Bernard R.: *The Astronomical Tables of Levi ben Gerson*. Transactions of the Connecticut Academy of Arts and Sciences, 45 (New Haven, 1974).

——: "Lunar Velocity in the Ptolemaic Tradition, in *The Investigation of Difficult Things: Essays on Newton and the History of the Exact Sciences*, ed. Peter M. Harman and Alan E. Shapiro (Cambridge, 1992), 3–17.

Goldstein, Bernard R., and José Chabás: "Ptolemy, Bianchini, and Copernicus: Tables for Planetary Latitudes," *Journal for the History of Astronomy*, 58 (2004), 453–473.

Hellman, Clarisse D., and Noel M. Swerdlow: "Peurbach (or Peuerbach), Georg," in *The Dictionary of Scientific Biography*, ed. Charles C. Gillispie, 16 vols. (New York, 1981), 15: 473–479.

Kennedy, Edward S.: *A Survey of Islamic Astronomical Tables*. Transactions of the American Philosophical Society, 46.2 (Philadelphia, 1956).

Lacher, Ambrosius: *Tabulae resolutae de motibus planetarum aliorumque super celestium mobilium* (Frankfurt, 1511).

Magrini, Silvio: "Joannes de Blanchinis Ferrariensis e il suo carteggio scientifico col Regiomontano (1463–1464)," *Atti e memorie della deputazione ferrarese di storia patria*, 22.3 (1917), 1–37.

Medica, Massimo: "Giovanni Bianchini, *Tabulae Astrologiae*: Miniato da Giorgio d'Alemagna," in *Le Muse e il principe: Arte di corte nel Rinascimento pagano*, 2 vols., ed. Alessandra Mottola Molfino and Mauro Natale, (Modena, 1991), 1: 186–189.

Mercier, Raymond: "Studies in the Medieval Conception of Precession (Part II)," *Archives Internationales d'Histoire des Sciences*, 27 (1977), 33–71.

Murr, Christoph Gottlieb von: *Memorabilia Bibliothecarum Publicarum Norimbergensium*, 3 vols. (Nuremberg, 1786), 1: 74–205.

Nallino, Carlo Alfonso: *Al-Battānī sive Albatenii Opus astronomicum*, 2 vols. (Milan, 1903–1907).

Neugebauer, Otto: *The Astronomical Tables of al-Khwārizmī* (Copenhagen, 1962).

——: "Thābit ben Qurra 'On the Solar Year' and 'On the Motion of the Eighth Sphere'," *Proceedings of the American Philosophical Society*, 106 (1962), 264–299.

——: *A History of Ancient Mathematical Astronomy* (Berlin, 1975).

North, John D.: "The Alfonsine Tables in England," in *Prismata: Festschrift für Willy Hartner*, ed. Yasukatsu Maeyama and Walter G. Saltzer (Wiesbaden, 1977), 269–301.

——: *Horoscopes and History* (London, 1986).

——: "Just whose were the Alfonsine Tables?," in *From Baghdad to Barcelona: Studies in the Islamic Exact Sciences in Honour of Prof. Juan Vernet*, ed. Josep Casulleras and Julio Samsó (Barcelona, 1996), 452–475.

Pedersen, Fritz S.: *The Toledan Tables: A Review of the Manuscripts and the Textual Versions with an Edition* (Copenhagen, 2002).

Porres, Beatriz: *Les tables astronomiques de Jean de Gmunden: édition et étude comparative*. Ecole pratique des hautes études, Section IV: Ph.D. dissertation (Paris, 2003).

Porres, Beatriz, and José Chabás: "John of Murs's *Tabulae permanentes* for Finding True Syzygies," *Journal for the History of Astronomy*, 32 (2001), 63–72.

Poulle, Emmanuel: "John of Murs," in *The Dictionary of Scientific Biography*, ed. Charles C. Gillispie, 16 vols. (New York, 1973), 7: 128–133.

——: *Les tables alphonsines avec les canons de Jean de Saxe* (Paris, 1984).

——: "The Alfonsine Tables and Alfonso X of Castile," *Journal for the History of Astronomy*, 19 (1988), 97–113.

Ratdolt, Erhard (ed.): *Tabule astronomice illustrissimi Alfontij regis castelle* (Venice, 1483).

Regiomontanus: *Epythoma Joannis de Monte Regio in Almagestum Ptolomei* (Venice, 1496).

Reinhold, Erasmus: *Prutenicae tabulae coelestium motuum* (Tübingen, 1551).

Rose, Paul L.: *The Italian Renaissance of Mathematics* (Geneva, 1975).

Rosińska, Grażyna: *Scientific Writings and Astronomical Tables in Cracow: A Census of Manuscript Sources (XIVth–XVIth Centuries)* (Wrocław, 1984).

——: "The 'Fifteenth-Century Roots' of Modern Mathematics," *Kwartalnik Historii Nauki i Techniki*, 41 (1996), 53–70.

——: "The 'Italian Algebra' in Latin and How it Spread to Central Europe: Giovanni Bianchini's *De Algebra* (ca. 1440)," *Organon*, 26–27 (1997), 131–145.

——: "The Euclidean *spatium* in Fifteenth-Century Mathematics," *Kwartalnik Historii Nauki i Techniki*, 43 (1998), 27–41.

——: "'Mathematics for Astronomy' at Universities in Copernicus' Time: Modern Attitudes toward Ancient Problems," in *Universities and Science in the Early Modern Period*, ed. Mordechai Feingold and Victor Navarro. Archimedes: New Studies in the History and Philosophy of Science and Technology, 12 (Dordrecht and Boston, 2006), 9–28.

Saby, Marie-Madeleine: *Les canons de Jean de Lignères sur les tables astronomiques de 1321*. Ecole Nationale des Chartes, Ph.D. dissertation (unpublished) (Paris, 1987). A summary appeared as: "Les canons de Jean de Lignères sur les tables astronomiques de 1321," *École Nationale des Chartes: Positions des thèses* (Paris, 1987), 183–190.

Santritter, Johannes Lucilius (ed.): *Tabule Astronomice Alfonsi Regis* (Venice, 1492).

Schöner, Johannes: *Tabulae Astronomicae, quas vulgo (...) resolutas vocant* (Nuremberg, 1536).

Steinschneider, Moritz: *Die hebraeischen Uebersetzungen des Mittelalters und die Juden als Dolmetscher* (Graz, [1893] 1956).

——: *Die hebraeischen Handschriften der k. Hof- und Staatsbibliothek in Muenchen* (Munich, 1895).

——: *Mathematik bei den Juden*, 2nd ed. (Hildesheim, 1964).

Suter, Heinrich: *Die astronomischen Tafeln des Muḥammad ibn Mūsā al-Khwārizmī* (Copenhagen, 1914).

Swerdlow, Noel M.: "Regiomontanus on the Critical Problems in Astronomy," in *Essays on Galileo and the History of Science in Honour of Stillman Drake*, ed. Trevor H. Levere and William R. Shea (Dordrecht, 1990), 165–195.

Thorndike, Lynn: "Giovanni Bianchini in Paris Manuscripts," *Scripta mathematica*, 16 (1950) 5–12 and 169–180.

——: "Giovanni Bianchini in Italian Manuscripts," *Scripta mathematica*, 19 (1953), 5–17.

Thorndike, Lynn, and Pearl Kibre: *A Catalogue of Incipits of Mediaeval Scientific Writings in Latin* (London, 1963).

Toomer, Gerald J.: "A Survey of the Toledan Tables," *Osiris*, 15 (1968), 5–174.

——: *Ptolemy's Almagest* (New York, 1984).

Zacut, Abraham: *Almanach Perpetuum* (Leiria, 1496).

Ziggelaar, August: "The Papal Bull of 1582 Promulgating a Reform of the Calendar," in *Gregorian Reform of the Calendar*, ed. George V. Coyne, Michael A. Hoskin, and Olaf Pedersen (Vatican, 1983), 201–239.

Zinner, Ernst: *Regiomontanus, his Life and Work*, translated by Ezra Brown (Amsterdam, 1990).

INDEX

History of Science and Medicine Library

ISSN 1872-0684

1. FRUTON, J.S. *Fermentation. Vital or Chemical Process?* 2006. ISBN 978 90 04 15268 7
2. PIETIKAINEN, P. *Neurosis and Modernity*. The Age of Nervousness in Sweden, 2007. ISBN 978 90 04 16075 0
3. ROOS, A. *The Salt of the Earth*. Natural Philosophy, Medicine, and Chymistry in England, 1650-1750. 2007. ISBN 978 90 04 16176 4
4. EASTWOOD, B.S. *Ordering the Heavens*. Roman Astronomy and Cosmology in the Carolingian Renaissance. 2007. ISBN 978 90 04 16186 3 (Published as Vol. 8 in the subseries *Medieval and Early Modern Science*)
5. LEU, U.B., R. KELLER & S. WEIDMANN. *Conrad Gessner's Private Library*. 2008. ISBN 978 90 04 16723 0
6. HOGENHUIS, L.A.H. *Cognition and Recognition: On the Origin of Movement*. Rademaker (1887-1957): A Biography. 2009. ISBN 978 90 04 16836 7
7. DAVIDS, C.A. *The Rise and Decline of Dutch Technological Leadership*. Technology, Economy and Culture in the Netherlands, 1350-1800 (2 vols.). 2008. ISBN 978 90 04 16865 7 (Published as Vol. 1 in the subseries *Knowledge Infrastructure and Knowledge Economy*)
8. GRELLARD, C. & A. ROBERT (EDS.). *Atomism in Late Medieval Philosophy and Theology*. 2009. ISBN 978 90 04 17217 3 (Published as Vol. 9 in the subseries *Medieval and Early Modern Science*)
9. FURDELL, E.L. *Fatal Thirst*. Diabetes in Britain until Insulin. 2009. ISBN 978 90 04 17250 0
10. STRANO, G., S. JOHNSTON, M. MINIATI & A. MORRISON-LOW (EDS.). *European Collections of Scientific Instruments, 1550-1750*. 2009. ISBN 978 90 04 17270 8 (Published as Vol. 1 in the subseries *Scientific Instruments and Collections*)
11. NOWACKI, H. & W. LEFÈVRE (Eds.). *Creating Shapes in Civil and Naval Architecture*. A Cross-Disciplinary Comparison. 2009. ISBN 978 90 04 17345 3
12. CHABÁS, J. & B.R. GOLDSTEIN (Eds.). *The Astronomical Tables of Giovanni Bianchini*. 2009. ISBN 978 90 04 17615 7 (Published as Vol. 10 in the subseries *Medieval and Early Modern Science*)

History of Science and Medicine Library

Medieval and Early Modern Science

Subseries Editors:
J.M.M.H. Thijssen and C.H. Lüthy

8. EASTWOOD, B.S. *Ordering the Heavens*. Roman Astronomy and Cosmology in the Carolingian Renaissance. 2007. ISBN 978 90 04 16186 3.
9. GRELLARD, C. & A. ROBERT (EDS.). *Atomism in Late Medieval Philosophy and Theology*. 2009. ISBN 978 90 04 17217 3
10. CHABÁS, J. & B.R. GOLDSTEIN (EDS.). *The Astronomical Tables of Giovanni Bianchini*. 2009. ISBN 978 90 04 17615 7

Published previously in the Medieval and Early Modern Science *book series*:

1. LÜTHY, C., J.E. MURDOCH & W.R. NEWMAN (EDS.). *Late Medieval and Early Modern Corpuscular Matter Theories*. 2001. ISBN 978 90 04 11516 3
2. THIJSSEN, J.M.M.H. & J. ZUPKO (EDS.). *Metaphysics and Natural Philosophy of John Buridan*. 2001. ISBN 978 90 04 11514 9
3. LEIJENHORST, C. *The Mechanization of Aristotelianism*. The Late Aristotelian Setting of Thomas Hobbes' Natural Philosophy. 2002. ISBN 978 90 04 11729 7
4. VANDEN BROECKE, S. *The Limits of Influence*. Pico, Louvain, and the Crisis of Renaissance Astrology. 2002. ISBN 978 90 04 13169 9
5. LEIJENHORST, C., C. LÜTHY & J.M.M.H. THIJSSEN (EDS.). *The Dynamics of Aristotelian Natural Philosophy from Antiquity to the Seventeenth Century*. 2002. ISBN 978 90 04 12240 6
6. FORRESTER, J.M. & J. HENRY (EDS.). *Jean Fernel's* On the Hidden Causes of Things. Forms, Souls, and Occult Diseases in Renaissance Medicine. 2005. ISBN 978 90 04 14128 5
7. BURTON, D. *Nicole Oresme's* De visione stellarum (On Seeing the Stars). A Critical Edition of Oresme's Treatise on Optics and Atmospheric Refraction, with an Introduction, Commentary, and English Translation. 2006. ISBN 978 90 04 15370 7

Printed in the United States
by Baker & Taylor Publisher Services